The Rational Use of Cognitive Resources

The Rational Use of Cognitive Resources

FALK LIEDER

FREDERICK CALLAWAY

THOMAS L. GRIFFITHS

PRINCETON UNIVERSITY PRESS

PRINCETON & OXFORD

Published by Princeton University Press
41 William Street, Princeton, New Jersey 08540
99 Banbury Road, Oxford OX2 6JX

press.princeton.edu

GPSR Authorized Representative: Easy Access System Europe - Mustamäe tee 50,
10621 Tallinn, Estonia, gpsr.requests@easproject.com

All Rights Reserved

ISBN 978-0-691-25995-6
ISBN (pbk.) 978-0-691-25994-9
ISBN (e-book) 978-0-691-25993-2
Library of Congress Control Number: 2025947054

British Library Cataloging-in-Publication Data is available

Editorial: Hallie Stebbins and Chloe Coy
Production Editorial: Jenny Wolkowicki
Cover design: Wanda España
Production: Lauren Reese
Publicity: William Pagdatoon
Copyeditor: Bhisham Bherwani

Cover image: James Thew / Adobe Stock

This book has been composed in ArnoPro

10 9 8 7 6 5 4 3 2 1

CONTENTS

1	Introduction	1
2	A Brief History of Rationality	16
3	The Rational Use of Cognitive Resources	36
4	(Ir)rationality Revisited	63
5	Strategy Selection, Metareasoning, and Learning to Be Rational	89
6	Strategy Discovery	111
7	Representations and Architectures	140
8	Improving Decisions	151
9	Conclusion	195
10	Appendix: Mathematical Details	205
Acknowledgments		233
Bibliography		235
Index		259

also address the broader group of psychologists, cognitive scientists, and neuroscientists trying to understand human minds and brains, as well as computer scientists interested in reproducing such systems in machines. We see these ideas as being relevant to this broader audience because understanding the rational use of cognitive resources offers not just a new way to think about the evidence for human irrationality in decision-making, but also a new perspective on many long-standing questions in cognitive science. Many theories in the social sciences and humanities are (implicitly or explicitly) built on assumptions about human rationality. This book provides constructive suggestions for how those assumptions can be improved. Moreover, the new approach to improving people's decisions that we present is relevant for anyone who wants certain decisions to be made well, including researchers, educators, and practitioners who care about improving people's lives or the functioning of groups, organizations, or society.

To understand the potential we see in the resource-rational approach, it is worth observing that the fundamental limitation of the classical notion of rationality is not just that it does not engage with how difficult it might be to make a decision, but that it does not engage with cognitive processes at all. Maximizing expected utility is fundamentally a **behaviorist** (Skinner, 1953) notion of rationality: it defines rational action purely in terms of the agent's response to its environment. By contrast, considering the question of how an agent should best deploy its cognitive resources to solve a problem places resource-rational analysis in firm contact with those cognitive processes. For the first time, we can say not just what action an agent should take but also how it should think about what to do. This is a powerful idea that extends across the psychologist's entire toolbox of cognitive processes—attention, planning, and memory all involve the use of limited resources, and resource-rational analysis gives us a new way to understand them.

Even more significantly, resource-rational analysis provides a new way to think about how to support and improve human cognition. When we use classical rationality to assess people's performance, we risk assuming that performance can be improved by teaching people the classically rational solutions to the problems they face. However, because these solutions are computationally costly, they may be impractical for many real-world decisions. Resource-rational analysis suggests an alternative approach: teaching people practical cognitive strategies that make better use of their limited cognitive resources. Thinking about computational costs also highlights another way that we can support human decision-making: building more computation into human environments. Using this approach, we can design **cognitive prostheses** to support longer-term decision-making (Lieder et al., 2019) and understand the circumstances under which behavioral **nudges** are most likely to be effective (Callaway et al., 2023).

In addition to a framework, tools, and examples, we hope to share with you our excitement about this approach and our vision for the many ways it can be used to understand and improve human minds. The cognitive revolution provided an alternative to behaviorism by showing how mathematics can be used to express rigorous hypotheses about cognitive processes (Miller, 2003). In the same way, we see resource-rational analysis as offering a path to a cognitive revolution in how we use and think about rationality when developing theories to explain, predict, and improve human behavior. The result is a blueprint for a new bridge between ideal solutions and cognitive processes that is directly informed by empirical data and supports interventions that meaningfully improve human lives. We hope this book will help you join us in building that bridge.

1.1   A simple example

Before we dive into any more details, we will give a simple example that illustrates what a resource-rational analysis looks like, how it differs from classical notions of rationality, and how this difference gives us new tools for understanding human behavior. This example takes the simplest kind of decision an agent could make, considers how it becomes more complicated when resources are limited, and shows a surprising connection to something people do that has long been considered irrational. It is also the first resource-rational analysis that any of us worked on (Vul et al., 2009).

Imagine you have a choice between two options. For simplicity, we will call them Option A and Option B. If you select Option A, you receive a reward with probability p. If you select Option B, you receive the same reward with a probability of $1 - p$. Which option should you choose?

From the perspective of classical rationality, this problem is trivial. If p is greater than 0.5, then you should choose Option A, otherwise choose Option B. But it becomes much more challenging if you don't know the value of p. In many real human decisions, assessing the probability of different outcomes is the hardest part. Doing so may require imagining future possibilities, considering different sources of evidence, recalling similar situations from memory, or engaging in various other cognitive processes.

To simplify things, we are going to model all of those different cognitive processes as methods of **sampling** a possible outcome. Whether we imagine, consider, or recall, we are getting a piece of information about the value of p. We will assume that obtaining this piece of information is equivalent to sampling a possible outcome for this decision—getting a result that favors Option A with probability p or Option B with probability $1 - p$.

Assuming these cognitive processes are costly—they take time and mental effort—making rational use of our cognitive resources in this setting is a matter of deciding how many samples to generate before choosing one of the options. Making this decision requires considering the cost of sampling relative to the cost of making an error. Finding the optimal trade-off between these two costs can be formulated as an optimization problem.

The solution to this optimization problem might be surprising: for a wide range of relative costs, it is optimal to take a single sample. Intuitively, there are few values of p where more samples are informative. If p is close to 1 or 0, then a single sample indicates the right answer with high probability. If p is close to 0.5, then more samples yield little benefit over choosing at random. If the cost of making an error is large relative to the cost of sampling, then the few cases where p falls in an intermediate range may be enough for it to make sense to draw more samples. But in many situations, one is enough.

This resource-rational analysis explains an aspect of human behavior that, on its surface, appears irrational. Given choices like this, where options are rewarded probabilistically, people tend to select those options with the probability that matches the probability of reward. This **probability matching** behavior is inconsistent with the classically rational solution, which is to always choose the option that is best according to all available information (Vulkan, 2000). However, probability matching makes sense if people are using a cognitive mechanism that behaves like sampling and have to pay a cost for each sample they generate.

While this is a simple example—perhaps the simplest we can imagine—it illustrates some themes that will reappear many times in this book. First, while the classically rational solution is easy to define abstractly, computing a concrete rational action is much more cumbersome and often impractical due to limited cognitive resources. Second, by taking people's limited cognitive resources into account, we can understand aspects of human behavior that differ systematically from the prescriptions of classical rationality. Finally, we can reach that understanding while abstracting away from the specific cognitive mechanisms involved. Throughout the book, we are going to define models at a fairly high level of generality, using processes such as sampling that are intended to stand in as abstract proxies for a variety of different cognitive mechanisms.

While we have not specifically explored ways of improving human decision-making in the simple kinds of decisions we are considering in this example, our resource-rational analysis of such decisions suggests some ways we could improve them. If we can identify situations where the cost of error is high relative to the cost of sampling, these would be good situations in which to signal that people might want to think carefully before responding. Likewise, if we have information about what range the value of p is likely to fall into, we might be able to tell people whether generating more than one sample is worthwhile. This is a way to support people's decisions by putting some of the computation they might otherwise have to perform into their decision environment.

In the rest of the book, we explore how making the move that we made here—going from a task to a decision about how to deploy our cognitive resources to best solve that task—can be used to shed light not just on decision-making, but also on planning, attention, and memory. The optimization problems involved in allocating these resources become increasingly complex, and we will present a framework for formulating and solving those problems. We will also show how the resulting models connect to traditional questions in cognitive psychology, and how they can be used to design effective interventions for agents with bounded cognitive resources.

1.2 Answers to skeptical questions

The work presented in this book is the culmination of a decade-long research program exploring applications of resource-rational analysis in cognitive science. Consequently, we have had the opportunity to present these ideas to various audiences and, hence, to field various skeptical questions from those audiences. To help you decide whether reading further is worth the cognitive resources involved, this section provides our answers to those skeptical questions. Of course, if you are already convinced, you can skip this section.

1.2.1 Why should I care?

The answer to this question depends on the person asking it. Depending on your interests, you may find different aspects of resource-rational analysis appealing.

If you are a **psychologist**, resource-rational analysis provides a different way to think about cognitive processes. Many of the phenomena that interest psychologists, such as problem-solving and deductive reasoning, are trivial from the perspective of classical rationality: the goal is simply to solve the problem. In these cases, the journey—the sequence of cognitive processes we engage in when solving the problem—is more important than the destination. For a long time, psychologists have developed theories about how people perform such tasks by manually assembling different cognitive mechanisms, strategies, or heuristics guided by empirical results. If you use resource-rational analysis, you can have computers do this tedious work for you: having defined a task and a set of elementary cognitive processes that could be involved, the optimal way to combine these processes can be derived automatically. This provides a way to answer questions about how people should think and what they should be thinking about, in the same way that classical rational analysis answers the question of how people should act. Consequently, resource-rational analysis makes deeper contact with the traditional preoccupations of cognitive psychology than rational analysis. It also provides a different way to design interventions to support human decision-making, focusing on guiding people toward better strategies and offloading some of the computation involved into the environment in which decisions are made.

If you are an **economist**, you are likely to be familiar with the idea of modeling human behavior as the solution to an optimization problem, but possibly also concerned that people systematically deviate from classical rationality. Resource-rational analysis is a way to create models of human behavior that are just as general—and use many of the same tools—as classical rational models, but also capture some systematic deviations from classical economic rationality. As we will discuss in the next chapter, this approach is aligned with recent work in economics, particularly in the **rational inattention** literature (Sims, 2003; Caplin and Dean, 2015), but also provides a way to more explicitly connect those ideas to cognitive processes, and thereby make the models' predictions even more accurate.

If you are a **neuroscientist**, having a theory of rational action that makes closer contact with cognitive mechanisms means that theory also makes closer contact with the brain. The successes of classical rational analysis led to a flurry of research on how the brain might implement probabilistic inference, typically by searching for one kind of computation or neural circuit that might solve this problem. The resource-rational perspective suggests that different instances of probabilistic inference might be solved in different ways, depending on the structure of the problem and the elementary cognitive processes available. For example, we will discuss a variety of sampling and planning algorithms that make sense to use under different circumstances. This provides a more nuanced set of targets for neuroscientists to search for in the brain. In addition, the core problem of selecting which cognitive processes to engage in has a structure that parallels that of **reinforcement learning**, creating the possibility that the extensive work on the neuroscience of reinforcement learning (Dolan and Dayan, 2013; O'Doherty et al., 2015; Lee et al., 2012; Mattar and Lengyel, 2022; Miller and Venditto, 2021) might also help us understand this metacognitive process. Moreover, resource-rational analysis is a promising vehicle for translating neuroscientific insights into the basic computations neurons and neural circuits perform and how costly those computations are into computational models of cognition and their resulting applications.

If you are a **computer scientist**, resource-rational analysis provides an opportunity to understand how human minds are so efficient in their use of limited cognitive resources to solve a wide range of problems and to define a formal framework for emulating this in machines. Bounded optimality is a notion that originated in the AI literature, and there are still a lot of opportunities to improve on the computational methods that are used to solve the problem of discovering efficient algorithms. Research in machine learning has explored a variety of ad hoc methods for deciding how much computation to perform (Graves, 2016; Banino et al., 2021). Formulating these problems more explicitly in terms of **rational metareasoning** (Russell and Wefald, 1991a) may be a way to improve on these methods. Finally, a formal framework that captures some of the ways that people systematically deviate from classical rationality is a valuable tool for making sense of human behavior. If you are interested in building systems that interact with people, then better understanding the mapping from people's preferences to their behavior is essential for being able to infer people's preferences from their behavior (Ho and Griffiths, 2022).

If you are a **curious person** who is primarily interested in how to make better decisions in your own life, this book offers a different perspective on what it actually means to be rational. Many books about decision-making emphasize human irrationality, focusing on deviations from classical rationality. Our perspective is more nuanced: when considering the resource constraints that human minds operate under, many of the seemingly irrational things people do make more sense. It is reasonable to want to be more rational in your own decisions, but the path to getting there has to respect those constraints. We discuss improving human decisions in detail in Chapter 8, which might be of particular interest to you. You can find more pragmatic advice in Tom's book *Algorithms to Live By* (Christian and Griffiths, 2016), which is not explicitly framed in terms of resource-rational analysis but shares with this book the idea that computer science has an important contribution to make to understanding rational action.

1.2.2   Isn't this just bounded rationality?

Immediately after everyone started to agree on how to define rationality, Herbert Simon pointed out that it was an unreasonable standard for human behavior (Simon, 1955). His concern was one we share: that humans have limited time and cognitive resources, making maximization of expected utility impractical. Simon wanted researchers to acknowledge the idea of **bounded rationality**—that the constraints under which human minds operate make classical rationality unachievable. This idea is very much a part of resource-rational analysis.

However, Simon viewed bounded rationality as an idea, rather than a formal framework. In a letter to the psychologist Gerd Gigerenzer, he wrote, "I have never thought of either bounded rationality or satisficing as precisely defined technical terms, but rather as signals to economists that they needed to pay attention to reality, and a suggestion of some ways in which they might. But I do agree that I have used bounded rationality as the generic term, to refer to all the limits that make a human being's problem spaces something quite different from the corresponding task environments: knowledge limits, computational limits, compatibility of component goals" (Augier and March, 2004, p. 406). This is where we differ: we view resource-rational analysis as a formal framework that can be useful for understanding the behavior of bounded agents, turning the idea of bounded rationality into a generalizable approach that can be used to derive precise predictions about cognition and behavior given a specification of the problem to be solved and the cognitive processes available for doing so.

In the same letter, Simon objected to the interpretation of bounded rationality in terms of constrained optimization—assuming that people solve an optimization problem similar to the one solved by a classically rational approach but with bounded resources providing constraints on the solution. In fact, he threatened to sue the next person who interpreted bounded rationality in this way. To anticipate any such litigation, we want to state that this is not how we think about resource-rational analysis. Rather, we see this approach as fundamentally changing the optimization problem implicit in the definition of rationality from one that is focused on external actions to one that is focused on internal computations. The simple example given above illustrates this: thinking about cognitive processes in terms of sampling leads to a completely new optimization problem, one that is focused on using those cognitive processes effectively.

It is notable that the one formal model that Simon published in his papers on bounded rationality has a very similar flavor to the models we consider in this book. Simon's model of **satisficing** (Simon, 1955) turns a choice problem into a stopping problem, where options are considered until one is found that is above a derived threshold. This kind of structure arises repeatedly in resource-rational models, where limited cognitive resources require us to consider our options (or gather information, or sample from distributions) sequentially. Many of the models we present later in the book focus on the problem of making intelligent decisions about when to stop thinking in a way that is very similar to satisficing.

1.2.3   How is this different from other
perspectives on rationality?

Understanding what it means to be rational, and how the definition of rationality relates to what people actually do, has been a preoccupation of many researchers from many different disciplines. We provide a detailed treatment of these ideas in Chapter 2, but want to briefly sketch how we see the relationship of resource-rational analysis to previous approaches.

The **heuristics and biases** research program carried out by Daniel Kahneman and Amos Tversky in the 1970s provided clear evidence that people's judgments and decisions deviate from classical rationality (Tversky and Kahneman, 1974; Kahneman et al., 1982). This research program was originally framed in terms of the computational costs of rationality: given that probabilistic inference and calculating expected utility are intractable, people must find some way to approximate them. Heuristics are the natural solution, and biases are the clues that we can use to identify those heuristics. We view resource-rational analysis as being very aligned with this original goal: considering how the costs of using cognitive resources trade off with performance gives us a way to characterize what makes a heuristic worth using. As a consequence, we can rederive some of the heuristics that Kahneman and Tversky discovered (see Chapter 4) and derive new heuristics directly from the formulation of the problem people have to solve (see Chapter 6).

The notion of **ecological rationality** was offered by Gerd Gigerenzer and colleagues as an alternative way of understanding heuristics (Gigerenzer and Todd, 1999a; Gigerenzer, 2000). Focusing on the interaction between cognition and the environment, ecological rationality explores the idea that particular heuristics might be adaptive solutions in particular environments. In this way, using heuristics should not be considered a deviation from rationality, but rather a reasonable strategy in itself. More recently, this perspective has been argued for from the perspective of the **bias-variance trade-off**, which shows that simple strategies can be best when information is limited (Gigerenzer and Brighton, 2009). Again, we see this approach as being well aligned with resource-rational analysis, and some of the heuristics identified by this research program fall out of our framework (see Chapter 6). However, one important difference is that we consider computational costs as an important additional factor that pushes people toward simpler heuristics, beyond the effect of the bias-variance trade-off.

Work on **rational inattention** in economics has explored how classical rationality can be brought into closer alignment with human behavior by assuming that gathering information is costly (Sims, 2003; Caplin and Dean, 2015). The resulting models consider how much investment an agent should make in gathering information—allocating attention to it—in order to find the best compromise between the costs it incurs and the benefits it provides for decision-making. This approach shares with the models we will discuss in this book the idea of introducing resource constraints into the formulation of the decisions that humans face. One difference is that the models we present here include cognitive processes—computations—as well as information gathering. However, we see the class of models developed under the flag of rational inattention as instances of resource-rational analysis.

Other researchers have used the term **computational rationality** to refer to applications of bounded optimality to cognitive science (Lewis et al., 2014; Gershman et al., 2015). We prefer to refer to these as resource-rational models (and have used this term since 2012; Lieder et al., 2012) as a way to emphasize the fact that they are about the rational use of cognitive resources and to avoid making a commitment to those resources being computational (something that allows this approach to make direct contact with rational inattention, for example).

1.2.4   Why are you proposing this now?

While the mathematical ideas behind bounded optimality were first proposed in the 1980s, it has been hard to use this approach to identify concrete cognitive strategies until the last decade or so. It has now become possible to actually define meaningful cognitive models based on bounded optimality because of innovations in the way we formulate and solve the underlying computational problems, and because of significant increases in the speed of the computers on which we solve those problems. Using the resulting models as the basis for designing effective behavioral interventions is potentially even more computationally challenging, which is one reason why it is a topic we have only been able to explore recently. One of the important things we do in this book is to provide you with the details on how we draw on ideas in artificial intelligence and statistics to set these problems up in a way that makes them (at least approximately) solvable.

1.2.5   So aren't you also creating a harder problem for human minds to solve?

Yes, we are defining a harder problem. Deciding how to think is certainly harder than deciding how to act. However, that harder problem does not necessarily need to be solved by human minds, and definitely does not need to be solved on the same timescale.

The problem of optimizing how we use our cognitive resources could be solved in different ways. One way would be simply considering all of the ways we could use our resources and choosing the best strategy. This is extremely computationally challenging—it involves presenting the harder version of the problem directly to human minds. But, importantly, there are other ways to solve this problem. One is to make use of the fact that we have repeated opportunities to perform particular tasks, and even new tasks often share components with tasks we have performed before. This means that we can *learn* better cognitive strategies over our lifetimes. Another is to recognize that variants of those same tasks were also faced by our ancestors, meaning that we can *evolve* better cognitive strategies across generations. In some cases, we do this via cultural evolution, coming up with intuitive strategies or algorithms we share with other people (e.g., Thompson et al., 2022). In other cases, biological evolution can act as the optimizing force—something that seems particularly plausible for making effective use of resources such as attention and memory that are shared with many other species (e.g., Tomlin et al., 2015).

In this way, although deciding how to think is harder than deciding how to act, humanity has had much more time to solve the problem. Individual human decisions are typically made on a timescale ranging from a few hundred milliseconds to a few weeks. Cognitive strategies are learned over a lifetime or tuned over many generations. And the way we use particular cognitive resources may pre-date all human minds.

We also expect that the mind solves the problem of deciding how to allocate its cognitive resources imperfectly. Resource-rational analysis is useful to the extent that people's cognitive strategies approximate the optimal solutions to these problems; it does not require that they be perfect. There is room for slippage, suboptimality, and satisficing in reasoning, learning, and evolution.

1.2.6   Doesn't this lead to an infinite regress?

Yes, we have introduced another level of abstraction into decision-making, and the same move can potentially be iterated. Thus, while our focus here is on deciding how to decide, we can imagine deciding how to decide how to decide, and so on. However, we do not see this kind of regress as a major problem. The key reason is that each level of abstraction has significantly diminishing returns. Deciding how to decide helps us explain a wide range of otherwise puzzling aspects of human behavior. Adding another level of abstraction—deciding how to decide how to decide—moves the focus from cognitive processes to cognitive architectures, which we explore in Chapter 7 (e.g., Milli et al., 2021). At this level, the question is how we should design a mind for it to be most effectively used by an agent that decides how to best allocate cognitive resources. While we can still make progress in exploring this question, the answers are fairly abstract and harder to relate directly to human behavior. Adding further levels of abstraction gets us even further from behavior and begins to engage us with questions that start to push us beyond the limits of cognitive science (e.g., how should you design an evolutionary process to select a cognitive architecture that allows an agent that optimally allocates cognitive resources to solve problems?). While these questions are theoretically interesting, we see the biggest payoffs of these ideas lying at the lower levels of the potentially infinite regress.[1]

1.2.7 Are you really not interested in whether people are actually rational?

Yes, really. Or, to be more precise, we are not interested in testing the hypothesis that people are actually rational under different criteria for rationality. To us, what matters more is whether a particular definition of rationality is useful for understanding human behavior. We use resource rationality as a **methodological assumption** that allows us to make progress in the broader project of making sense of people's thoughts and actions (Godfrey-Smith, 2001). Classical rationality is arguably useful in this respect, even if people deviate from it in systematic ways. Most Bayesian models of cognition are implicitly formulated within this framework. These models help us understand the aspects of people's behavior that we can explain in terms of the optimal solutions to the problems that human minds face (Anderson, 1990; Griffiths et al., 2010). Resource-rational analysis allows us to go a step beyond that, introducing assumptions about constrained cognitive resources and using them to explain some of the systematic deviations from classical rationality.

Given the vast space of possible Bayesian models and cognitive mechanisms, we anticipate (and our critics have suggested; Jones and Love, 2011; Bowers and Davis, 2012) that it will be hard to falsify the assumption of either of these forms of rationality. For that reason, we do not think it is productive to engage in debates about whether or not people are rational. Rationality, of whatever denomination, provides a *framework* in which we can develop *models* of human behavior (Lakatos, 1970). Individual models are falsifiable—they generate specific predictions that we can test in further experiments.

---

1. It is worth noting that there are also precedents of not being too concerned about infinite regress in the context of understanding decision-making. The classic example is in game theory, which considers the asymptotic equilibria of processes where players deliberate about the consequences of their actions. Whether similar asymptotic results can be obtained in the case of resource-rational analysis is an interesting research question.

Establishing whether people's behavior is consistent with specific models allows us to evaluate whether the factors considered in those models—hypothesis spaces and prior distributions in Bayesian models, augmented by cognitive mechanisms and constraints in resource-rational models—are candidates for explaining different aspects of that behavior.

Ultimately, we take the success of specific models based on resource-rational analysis as evidence that the framework is *useful*, not as evidence that the assumption that people always act in a resource-rational way is *true*. The approach we present in this book will be successful if it helps us develop theories that help us predict and explain why people act in the way they do and design interventions that are effective in improving that behavior.

1.3 The rest of the book

If you have gotten this far, you are hopefully satisfied with our answers to your skeptical questions—or are at least reserving judgment until you get more details. In the same spirit of letting you make an informed decision about how to use your time, we will use the remainder of this chapter to briefly summarize the rest of the book. Our focus in the book is on presenting the theoretical ideas behind resource-rational analysis and illustrating these ideas through some of the applications and experimental results that demonstrate the value of this approach. However, this is still a new perspective and we anticipate that some of the strongest evidence for its utility is still to come. We aim to equip readers to be able to gather that evidence for themselves.

Chapter 2 provides a brief history of different perspectives on rationality in psychology, economics, and computer science, expanding on the description that appears above and providing more details on the background behind resource-rational analysis and the context in which it arose.

Chapter 3 introduces the key ideas behind the formal framework of resource-rational analysis, drawing on bounded optimality and rational metar-easoning to define a set of mathematical tools that we will use to define problems and state their solutions in the rest of the book.

Chapter 4 revisits the evidence for human irrationality, providing an analysis of some of the classic heuristics proposed by Kahneman and Tversky from the perspective of resource-rational analysis. The heuristics include **anchoring-and-adjustment** (Tversky and Kahneman, 1974) and the **availability heuristic** (Tversky and Kahneman, 1973), both of which are related to ideas about how to best use sampling algorithms to estimate probability distributions.

Chapter 5 applies resource-rational analysis to strategy selection. According to explanations of human decision-making postulating heuristics, the mind is equipped with a toolbox full of different decision strategies. This raises the question of how people know when to use which strategy. This chapter explains how the problem of selecting a strategy can be expressed as a form of rational metareasoning—a problem that has been studied in the artificial intelligence literature (Russell and Wefald, 1991a). We then present a theory according to which people solve this problem by learning to predict how well different strategies work in different situations. We conclude the chapter with some empirical findings demonstrating that people can learn to become more rational.

Chapter 6 turns from strategy selection to strategy discovery. While strategy selection is about choosing between known strategies, strategy discovery is much harder: it requires us to put together new strategies out of elementary cognitive operations. In this chapter, we show how this problem can be solved using **metalevel Markov decision processes** (Hay et al., 2012), with applications to decision-making, planning, and memory.

Chapter 7 considers the implications of resource-rational analysis for two other key questions in cognitive science: how should we represent the world and how should human minds be structured? In answering the first question, we show how thinking about representations in terms of the computational costs they incur can account for how people form abstractions (Ho et al., 2022; Correa et al., 2023). In answering the second question, we show how resource rationality favors minds composed of a small number of cognitive systems (justifying the idea of having a "fast" and a "slow" system) (Milli et al., 2021).

Chapter 8 considers how resource-rational analysis can be used as a guide to constructing effective interventions for improving people's decisions. In this chapter, we show how having a more accurate normative standard for people's behavior can allow us to better identify real opportunities for improvement and to derive effective interventions to achieve those improvements. In particular, we discuss the approaches we have taken to teach people resource-rational decision strategies (Callaway et al., 2022a), construct cognitive prostheses (Lieder et al., 2019), and nudge people toward making better decisions with less effort (Callaway et al., 2023).

Chapter 9 concludes the book, summarizing the results from the previous chapters and considering their implications, as well as highlighting future directions for this research program.

We have constructed the book with the intent that it be read from start to finish, with later chapters building on ideas presented in earlier chapters and assuming some of their content as context. Hence, perhaps appropriately, the main decision you have to make about how to best use your cognitive resources is when to stop reading and start acting on the things you have learned.

2

A Brief History of Rationality

People have been wondering how to make good decisions for hundreds of years. In the seventeenth century, the mathematician Blaise Pascal considered the question of how to decide whether or not to believe in God (Pascal, 1702). He framed it as a wager, taking into account the probability that God exists or does not exist and the rewards and costs associated with belief in both cases. He analyzed this decision by calculating the average value of each action—the sum of the rewards multiplied by the probabilities for each outcome. Assuming the rewards associated with believing in a God who exists are infinite, he argued that belief is worthwhile even if the probability of God's existence is small.

In making this argument Pascal made two bold assumptions—that we should *average* over outcomes, and that we can directly assess the *value* of those outcomes. This notion of making decisions based on the **expected value** of different actions came under fire from the cousins Nicolas and Daniel Bernoulli. Nicolas showed that considering expected value could lead to a paradox (the original version appears in a letter to Pierre de Montmort reproduced in de Montmort, 1713).

Consider the following game: $1 is put in front of you, and you flip a coin. If the coin comes up tails, the money is doubled and the process repeats. When the coin eventually comes up heads, you take the money. How much should you be willing to pay to play this game? The expected value is $\frac{1}{2} \times 1 + \frac{1}{4} \times 2 + \frac{1}{8} \times 4 \ldots = \frac{1}{2} + \frac{1}{2} + \frac{1}{2} \ldots$, which sums to infinity. So, in principle, you should be willing to pay an arbitrarily large amount of money to play this game—something that is not what most people would choose.

Daniel Bernoulli resolved this puzzle, which came to be called the **St. Petersburg paradox**. Instead of focusing on the monetary value of each outcome, his solution to the puzzle postulates that people assign a **subjective value** to that outcome (Bernoulli, 1954). If the subjective value of x increases more slowly than x, then the paradox can be avoided. In particular, he proposed taking the subjective value to be $\log x$, in which case the infinite sum converges. Nicolas Bernoulli had another solution, which is that people might assign a **subjective probability** to different outcomes, discounting small probabilities (de Montmort, 1713).

These two ideas—subjective value and subjective probability—would come to play a large role in efforts to formalize and critique rational decision-making in the twentieth century. In this chapter, we briefly review these efforts, starting with work that explored the formal foundations of decision-making and then considering a series of critiques that arose as that work came into contact with human behavior. These ideas provide the foundation for resource-rational analysis, which we introduce in detail in the next chapter.

2.1 Formalizing rational decision-making

Interest in understanding rational decision-making grew in the twentieth century as the disciplines of economics and probability theory matured. Efforts to precisely define how people should take into account uncertainty when making decisions were made by Frank Ramsey (Ramsey, 1990) and John Maynard Keynes (Keynes, 1936), but the canonical characterization of rational decision-making is that given by John von Neumann and Oskar Morgenstern in their 1944 book *Theory of Games and Economic Behavior* (von Neumann and Morgenstern, 1944).

Von Neumann was a mathematician in the process of transitioning from being preoccupied with the formal foundations of mathematics to becoming one of the pioneers of electronic computing. Morgenstern was an economist just starting to recognize how formalization could help provide a new foundation for his discipline. They spent time together at the Institute for Advanced Study in Princeton, and the outcome was a book that set much of the agenda for the study of decision-making for the remainder of the twentieth century.

The approach that von Neumann and Morgenstern took was one that had been successful in establishing the foundations of other branches of mathematics: define a set of **axioms** that describe the object of interest and then demonstrate their consequences. In the case of decision-making, the basic object was **preferences**: that a person might prefer one outcome to another. Given outcomes A, B, C, and D, we can say somebody prefers A to B (written $A \succ B$) if they would be happier if A occurred. We can likewise say that they are indifferent between B and C ($B \sim C$) and that they do not disprefer C to D ($C \succsim D$).

To illustrate what the relevant axioms look like, we will state them briefly, but we will not make further use of these axioms in the remainder of the book. Von Neumann and Morgenstern were interested in identifying a set of principles that could uncontroversially characterize how a rational agent should act. Their focus was on simple gambles, where different outcomes occurred with different probabilities. For example, a gamble in which A occurs with probability p and B occurs with probability $(1 - p)$ could be written $\{pA, (1 - p)B\}$. In simple terms, the four key axioms were:

1. **Completeness**: For any two options A and B either $A \succsim B$ or $B \succsim A$ (or both).
2. **Transitivity**: If $A \succsim B$ and $B \succsim C$ then $A \succsim C$.
3. **Continuity**: If $A \succsim B \succsim C$, then there exists some p such that $\{pA, (1 - p)C\} \sim B$.
4. **Independence of irrelevant alternatives**: If $A \succsim B$ then for any p it must be true that $\{pA, (1 - p)C\} \succsim \{pB, (1 - p)C\}$.

From these axioms, von Neumann and Morgenstern proved a striking result: that if and only if the preferences of a decision-maker follow these axioms, then those preferences can be captured by a **utility function** $U(\cdot)$ mapping outcomes to real numbers such that $U(A) \geq U(B)$ if and only if $A \succsim B$. Furthermore, the preferences among lotteries can be computed by calculating the **expected utility** of those lotteries, with $U(\{pA, (1 - p)B\})$ being $pU(A) + (1 - p)U(B)$. An agent whose preferences obey von Neumann and Morgenstern's axioms would thus choose actions based on the expected utility of their outcomes, justifying the assumptions that previous mathematicians had made about the nature of rational decision-making.

The idea of a utility function provides a way to capture the notion that people might assign a subjective value to different outcomes. The same kind of approach was used to extend this analysis to subjective probabilities, allowing for cases where people make decisions based on their own assessment of the odds rather than the objective probabilities in the lotteries considered by von Neumann and Morgenstern. Building on the approach of Ramsey, Leonard Savage (Savage, 1954) and Richard Jeffrey (Jeffrey, 1965) came up with similar axiomatizations of decision-making under uncertainty from which the properties of subjective probability functions as well as utility functions can be derived.

This methodology of starting from a reasonable set of axioms and deriving their consequences provided the foundations of theories of decision-making in statistics and economics throughout the twentieth century. While decision theorists continue to refine these axioms and consider how notions of rational action might be modified within this kind of formal framework, the idea of rational choice as maximizing expected utility remains the standard against which human decisions are often assessed.

Part of the appeal of the idea of maximizing expected utility as a model of decision-making is its generality. Given a set of outcomes, their probabilities, and a utility function that summarizes a person's preferences, it is straightforward to predict the decisions that person will make. "Soft" versions of maximizing expected utility, in which people are assumed to select actions with a probability that increases with their relative expected utility, can be defined by using the "softmax" rule,

$$P(\text{choose } A) \propto \exp\{\beta U(A)\}, \tag{2.1}$$

where $\beta = 0$ results in random choice and $\beta \to \infty$ converges to only choosing options that have maximum utility. This assumption is widely used for making inferences about people's preferences from their behavior in both economics and computer science (e.g., McFadden, 2001; Dragan et al., 2013).

2.2 Bounded rationality

Despite its simplicity and generality, maximizing expected utility quickly came under criticism as an account of human decision-making. In two papers, published in 1955 and 1956, Herbert Simon argued strongly against the assumptions underlying these models, and equally strongly for the idea that accounts of human decision-making should take into account the real constraints under which human minds operate (Simon, 1955, 1956). This alternative perspective is known as **bounded rationality**.

Herbert Simon had trained as a political scientist and economist, and his early work had focused on how organizations operate. He would go on to become one of the founders of cognitive science and artificial intelligence, developing the first computer models of human cognition (Newell and Simon, 1956) and doing seminal work on understanding human problem-solving (Newell et al., 1972)—ideas that we will return to later in the book. He later received a Nobel Prize for his work studying decision-making processes in organizations and a Turing Award for his foundational work in AI.

In his papers on bounded rationality, Simon highlighted the unrealistic assumptions that go into the axiomatic analyses summarized above. The first paper targeted economists, and took on axiomatic treatments of rational decision-making directly. Describing the hypothetical "economic" or "rational man" assumed in these accounts, he wrote:

> This man is assumed to have knowledge of the relevant aspects of his environment which, if not absolutely complete, is at least impressively clear and voluminous. He is assumed also to have a well-organized and stable system of preferences, and a skill in computation that enables him to calculate, for the alternative courses of actions that are available to him, which of these will permit him to reach the highest attainable point on his preference scale. (Simon, 1955, p. 99)

To Simon, based on his experience studying real human decision-making in real organizations, these assumptions seemed implausible.

Simon sought to present an alternative. He went on to write, "Broadly stated, the task is to replace the global rationality of economic man with a kind of rational behavior that is compatible with the access to information and the computational capacities that are actually possessed by organisms, including man, in the kinds of environments in which such organisms exist" (p. 99). Interestingly, he did not consider psychology to be mature enough for this task, noting the great distance between "our present psychological knowledge of the learning and choice processes and the kinds of knowledge needed for economic and administrative theory" (p. 100), so instead he focused on identifying some of the ways in which people might differ in their decision-making from the idealized version that appears in axiomatic theories of choice.

Simon's emphasis was on the constraints faced by real human decision-makers. Instead of just thinking about an abstract agent in an abstract environment, we need to consider the cognitive limits of the agent and how it interacts with that environment. However, he also noted that "we must be prepared to accept the possibility that what we call 'the environment' may lie, in part, within the skin of the biological organism" (p. 101). In doing so, he put the emphasis on the biological and cognitive limits that human minds are subject to, making particular note that "limits in computational capacity may be important constraints entering into the definition of rational choice under particular circumstances" (p. 101).

Simon supported this general perspective with specific examples, arguing that decision-makers might not have strong preferences about all possible outcomes and might simply be willing to accept an outcome that is good enough, and might not have the opportunity to exhaustively explore all possibilities. This leads to an alternative view of decision-making, in which people **satisfice**—they consider options until they find one that is good enough, and then act. Simon presented a simple version of such a satisficing model, in which a decision-maker considers a series of offers until one exceeds a threshold. This model turns decision-making into an **optimal stopping** problem—deciding on a threshold to apply that is a good compromise between the ultimate payoff and the effort that has gone into considering the options. Later in the book, we will present models that apply a very similar framework to understanding a variety of questions about how to make the best use of limited cognitive resources.

Simon's second paper was directed at psychologists, and emphasizes the role of the environment in guiding what constitutes adaptive behavior (Simon, 1956). Simon's argument was that "if an organism is confronted with the problem of behaving approximately rationally, or adaptively, in a particular environment, the kinds of simplifications that are suitable may depend not only on the characteristics—sensory, neural, and other—of the organism, but equally upon the structure of the environment" (pp. 129–130). He went on to show that, in a simple environment, a simple strategy can result in adaptive behavior. Thus, the complex machinery of axioms and utility functions may provide less insight into what the organism is doing than the constraints that are provided by the environment in which it operates. Consequently "we should be skeptical in postulating for humans, or other organisms, elaborate mechanisms for choosing among diverse needs" (p. 137).

These two factors—constraints on the agent and the nature of the environment—are at the heart of the way that Simon thought about bounded rationality. He would later characterize human behavior being shaped by "scissors whose two blades are the structure of task environments and the computational capabilities of the actor" (Simon, 1990, p. 7). This leaves an interesting challenge: determining exactly how human behavior is shaped by those factors. We view the framework we present in this book as a way of taking up that challenge, providing tools for answering the question of how environments and computational constraints *should* shape an agent as a precursor to understanding how they *do* shape humans. In this way our goal is to understand what Simon called "the approximating procedures people use in many kinds of complex situations," seeking "to characterize these procedures, to show how they are acquired, and to account for their compatibility with the known computational limitations of the human brain" (Simon, 1990, p. 8).

Consistent with our strategy of first identifying how the constraints should shape people's cognitive strategies, Simon suggested that knowledge about people's boundedness can be integrated into theories of rationality (Simon, 1978). In the 1970s and 1980s, several psychologists took up the task of investigating which approximate decision procedures might be rational for real people given that their computational capacities are bounded (e.g., Beach & Mitchell, 1978; Christensen-Szalanski, 1978; Russo & Dosher, 1983; Payne, Bettman, & Johnson, 1988). Their research compared the costs and benefits of using various heuristics. Along similar lines, Baron (1985) defined rational thinking as the use of a strategy that achieves the highest expected utility when the cost of thinking is taken into account. According to his theory, we should not expect people always to use the strategy that is optimal for the exact situation they are in. Instead, people can only be expected to categorize situations and then use the rational strategy for situations of that type. Baron (1985) then proceeded to outline how these ideas could be formalized by applying decision theory to the selection of cognitive operations.

Table 2.1. The Allais gambles: Participants choose between lottery L_1 and lottery L_2 for $z = 2{,}400$ versus $z = 0$. Each lottery is described by a set of pairs of outcomes o_i and their probabilities p_i.

	$L_1(z)$	$L_2(z)$
(o_1, p_1)	$(z, 0.66)$	$(z, 0.66)$
(o_2, p_2)	$(2{,}500, 0.33)$	$(2{,}400, 0.34)$
(o_3, p_3)	$(0, 0.01)$	

2.3 Heuristics and biases

Simon's objections to the classical definition of rational decision-making was primarily theoretical, although motivated by his experience studying how humans made decisions within organizations. These objections were quickly complemented by empirical demonstrations of ways in which human decision-making systematically deviated from the prescription of maximizing expected utility.

One early example of this is what is known as the **Allais paradox**. Published by the economist Maurice Allais in 1953, it provided a simple example of a case where people's intuitions seemed to violate the axiom of independence of irrelevant alternatives (Allais, 1953). In the two lotteries $L_1(z)$ and $L_2(z)$ defined in Table 2.1, the chance of winning z dollars is exactly the same. Yet, when $z = 2{,}400$ most people prefer lottery L_2 over lottery L_1, but when $z = 0$, the same people prefer L_1 over L_2.

Further empirical evidence that people violate the axioms used to characterize rationality came from the work of Daniel Kahneman and Amos Tversky.[1] Both were trained as psychologists, but they came from quite different traditions. Kahneman had worked extensively on human perception and attention, and the measurement of mental effort. Tversky had a background in mathematical psychology and decision theory, taking an axiomatic approach to modeling different aspects of human cognition. Together, they launched an extremely influential research program exploring the ways in which people deviate from classical rationality, culminating in a Nobel Prize.

Kahneman and Tversky wrote a series of papers together in the early 1970s that documented systematic differences between the way that people are

---

1. We will maintain this order of names throughout this section for simplicity, but Kahneman and Tversky contributed equally to this work and alternated authorship on the relevant papers.

supposed to judge probabilities and estimate quantities and the way they actually solve these problems. In their 1974 paper in *Science* that summarized these results, they stated their thesis as follows:

> This article shows that people rely on a limited number of heuristic principles which reduce the complex tasks of assessing probabilities and predicting values to simpler judgmental operations. In general, these heuristics are quite useful, but sometimes they lead to severe and systematic errors. (Tversky and Kahneman, 1974, p. 1124)

Their starting point in this statement is the high computational cost of probabilistic inference, which motivates the need for simpler solutions. Those simpler solutions—**heuristics**—result in systematic errors—**biases**. These biases are a source of clues about the heuristics people use. The result became known as the **heuristics and biases** research program (e.g., Kahneman et al., 1982).

We will explore some of the heuristics discovered by Kahneman and Tversky in detail later in the book, so it is worth taking a moment to briefly consider some examples.

The **representativeness heuristic** involves replacing a probability judgment with one of similarity, on the assumption that similarity is more cognitively accessible (Kahneman and Tversky, 1972). As an example, representativeness could be appealed to to explain why people consider a sequence of heads (H) and tails (T) such as HHTHT more likely to be produced by a fair coin than HHHHH. Even though the two sequences are in fact equally probable, the first contains a more balanced number of heads and tails and is hence more representative.

The representativeness heuristic can lead to systematic errors in probability judgment. For example, if we follow the axioms of probability theory, it is impossible for the conjunction of two propositions to have a higher probability than one proposition alone. However, considering representativeness can lead people to violate this—a phenomenon known as the **conjunction fallacy** (Tversky and Kahneman, 1983). For example, consider the following description:

> Linda is 31 years old, single, outspoken and very bright. She majored in philosophy. As a student, she was deeply concerned with issues of discrimination and social justice, and also participated in antinuclear demonstrations.

More than 80% of undergraduates surveyed thought it was more probable that Linda was a bank teller and active in the feminist movement than that Linda was a bank teller (Tversky and Kahneman, 1983).

The **availability heuristic** uses another cognitive quantity as a proxy for probability: the ease with which something is imagined or recalled from memory (Tversky and Kahneman, 1973). For example, in one of Kahneman and Tversky's experiments, participants thought that there were more words that began with the letter R than had R as the third letter, despite the truth being the reverse. This is because it is easier to think of examples of words that begin with a particular letter than words that have that letter within them.

Finally, Kahneman and Tversky found that people sometimes make judgments via a process of **anchoring-and-adjustment**: they begin with an initial estimate and then modify that estimate based on further reflection and internal cues from memory and reasoning (Tversky and Kahneman, 1974). In a classic experiment, participants were asked to estimate the proportion of the countries in the United Nations that are from Africa. They first saw a number between 1 and 100 generated by spinning a wheel, and had to judge whether the true number was higher or lower; then, they stated their estimate. This initial number—the anchor—influences people's estimates: participants who saw 10 gave a median estimate of 25, while those who saw 65 gave a median estimate of 45. They thus insufficiently adjusted their estimates away from the anchor.

Despite Tversky's background in mathematical psychology and decision theory, Kahneman and Tversky's analysis of these heuristics was not particularly formal. That mathematical expertise primarily went into identifying the "weak spots" of existing accounts of rational action, focusing on violations of key axioms and predictions. One notable exception, however, is **prospect theory**, the model of decision-making that Kahneman and Tversky developed using a similar approach to their work on heuristics and biases (Kahneman and Tversky, 1979).

Applying the same lens to decision-making that they had to judgment, Kahneman and Tverksy identified several meaningful ways in which people's intuitions do not align with maximizing expected utility, including the Allais paradox. They then proposed prospect theory as a way to better capture these intuitions. Their basic assumption was that people would value a lottery via a functional form similar to that of expected utility, but with probabilities and values applied to the options that systematically deviate from the objective probabilities and values. Specifically, the value V assigned to the lottery $\{pA, (1-p)B\}$ would be $\pi(p)V(A) + \pi(1-p)V(B)$, where $\pi(\cdot)$ is a function transforming probabilities into weights and $V(\cdot)$ is a subjective value function. Looking back at the proposed solutions to the St. Petersburg paradox from the beginning of this chapter, it seems that *both* of the Bernoulli cousins were onto something.

While using subjective probability weights and value functions may seem like a subtle change, it made it possible to capture systematic patterns in human decision-making. The probability weighting function $\pi(\cdot)$ was taken to overweight small probabilities and underweight larger ones. The value function $V(\cdot)$, unlike the utility function $U(\cdot)$, allowed for the context in which a prospect was being valued, with losses being treated differently from gains. These two changes were sufficient to reproduce all of the phenomena motivating Kahneman and Tversky's account. For example, the Allais paradox can be explained if people overweight small probabilities.

Kahneman and Tversky's work resulted in a substantial literature documenting the heuristics that people follow and the biases that result from them (see, for example, Kahneman et al., 1982; Gilovich et al., 2002). This perspective spread from psychology into economics, resulting in the development of behavioral economics. However, this work leaves some important questions unanswered: What makes something a good heuristic? Why are these heuristics the ones that human minds use? We hope to be able to answer these questions by applying a more formal analysis to understanding how people reduce the complex tasks of assessing probabilities and predicting values to simpler judgmental operations.

2.4 Ecological rationality

Much of the literature on heuristics and biases that has followed Kahneman and Tversky's work has focused on the biases rather than the heuristics. Consulting popular books about decision-making, we might learn that we are "predictably irrational" (Ariely, 2009), and that our minds are based on "kluges" (Marcus, 2008) that result in "inevitable illusions" (Piattelli-Palmarini, 1994). An alternative perspective is offered by the work of Gerd Gigerenzer and his colleagues, who have considered the possibility that the heuristics we use are adaptive for the environments in which we often make decisions (Gigerenzer and Todd, 1999b).

Gigerenzer's engagement with heuristics and biases began with the observation that when people are presented with problems of probabilistic reasoning where the quantities involved are presented as *frequencies* rather than *probabilities*, they perform better. For example, if the description of Linda given above is followed by a judgment of frequencies—

There are 100 persons who fit the description above (i.e., Linda's). How many of them are:
(a) bank tellers
(b) bank tellers and active in the feminist movement?

—then the proportion of participants making the error goes down to just 22% (Fiedler, 1988). Gigerenzer argued that a similar reduction in errors occurs for other probabilistic reasoning tasks (Gigerenzer, 1991;

Gigerenzer and Hoffrage, 1995). A response to this argument appears in Kahneman and Tversky (1996).

A second line of work explored settings where heuristics can be extremely effective. For example, consider a decision-making problem that requires considering options that have many different features. Focusing on the one feature that best predicts the outcome is a way to reduce the complexity of the decision. It turns out that this **take the best** heuristic is also extremely effective in many naturalistic decision problems (Gigerenzer and Goldstein, 1996). In fact, it can also be shown to be the optimal strategy in **non compensatory** environments, where the other features do not contribute enough information to compete with the most predictive feature (Gigerenzer and Goldstein, 1999; Martignon and Hoffrage, 1999).

These two lines of work led Gigerenzer and colleagues to the idea of **ecological rationality**: "A heuristic is ecologically rational to the degree that it is adapted to the structure of an environment" (Gigerenzer and Todd, 1999a, p. 13). The emphasis on the role of the environment in determining what kinds of strategies might be reasonable for an agent to adopt explicitly reinvokes one of the two blades of the scissors that Simon viewed as shaping human behavior. The other blade, computational capacity, is reflected in the expectation that these heuristics be relatively simple. Or, as they put it, "Fast and frugal heuristics that are matched to particular environmental structures allow organisms to be ecologically rational" (Gigerenzer and Todd, 1999a, p. 18).

Early work on ecological rationality would proceed by identifying a heuristic, establishing that people seem to use it, and characterizing the environments in which that heuristic is adaptive (for examples, see Gigerenzer and Todd, 1999b). More recently, the preference for fast and frugal heurstics has been explained in terms of the **bias-variance trade-off** from statistics (Gigerenzer and Brighton, 2009). This classic result is based on analyzing the prediction performance of learning algorithms (Geman et al., 1992). There are two ways to make bad predictions: use a model that is too simple, in which case the model has a bias that is not aligned with the data; or use a model that is too complex, in which case the predictions vary significantly based on idiosyncratic pseudo-patterns (or "noise") in the data. The bias-variance trade-off is the idea that, as model complexity increases, we trade bias for variance, so we should try to find a sweet spot where neither is too large. In particular, since variance increases when samples are small, when making predictions from limited data we should be willing to accept a little bias. The bias-variance trade-off thus justifies using simple heuristics when the environment doesn't give us a lot of information on which to make predictions.

Ecological rationality provides one answer to the question of what makes a good heuristic. However, there are still important ways in which this answer is incomplete, and other questions that it raises. First, is minimizing variance the only reason why we should prefer simple heuristics? This says nothing about the computational cost of different strategies we employ, which could make us prefer simplicity even when the environment gives us plenty of information. Second, if we have a toolbox of fast and frugal heuristics, how do we decide which tool to use in a given environment? And finally, and perhaps most importantly, how do we—as scientists, but also as adaptive agents—discover effective heuristics? We will engage with all three of these questions later in the book.

2.5 Rational analysis

Kahneman and Tversky's work had given rationality a bad name among psychologists, but in his book *The Adaptive Character of Thought*, John Anderson suggested that it might still have value as a framework for understanding human cognition (Anderson, 1990). Inspired by Marr (1982) and Shepard (1987), Anderson argued that some of the puzzling things that people do could be explained by considering the abstract computational problems that human minds need to solve, and assuming that those problems are grounded in the environments in which those minds live. This key role of the environment in defining what a rational solution might look like echoes the argument made by Simon (1956).

Anderson's approach of **rational analysis** proceeds in a series of five steps.

1. Precisely specify what are the goals of the cognitive system.
2. Develop a formal model of the environment to which the system is adapted.
3. Make minimal assumptions about computational limitations.
4. Derive the optimal behavioral function given items 1 through 3.
5. Examine the empirical literature to see if the predictions of the behavioral function are confirmed.

If the resulting predictions do not match the data, then the process can be iterated.

While this approach offers a different way of thinking about some results in judgment and decision-making, the real impact of rational analysis was as a framework for characterizing optimal solutions to other kinds of computational problems people face—problems that make contact with classic questions in cognitive psychology about how people learn categories (Anderson, 1991; Sanborn et al., 2010), organize information in memory

(Anderson and Milson, 1989; Anderson and Schooler, 1991), infer causal relationships (Anderson and Sheu, 1995; Griffiths and Tenenbaum, 2005), and reason about the world (Oaksford and Chater, 1994; Griffiths and Tenenbaum, 2006). Many of these questions relate to **inductive problems**, where the available evidence underconstrains the conclusion. For example, learning categories requires going from examples of members of those categories to judgments about whether other objects belong to those categories. For these problems, the ideal solution comes in the form of **Bayesian inference** (for a history of and an introduction to the Bayesian way of thinking, see Jaynes, 2003).

Just as axiomatic treatments of decision-making had shown that rational agents should maximize expected utility, a variety of mathematical arguments show that rational agents should represent their degrees of belief about the world via probabilities. That is, if we want to avoid being out-gambled by our competitors (the "Dutch book" argument; Ramsey, 1990; de Finetti, 1964) or obey some intuitive principles of inductive inference (Cox's theorem; Cox, 1946, 1961), we should use probabilities to encode our degrees of belief.

If we accept the idea that degrees of belief should be encoded with probabilities, then the way we should update those degrees of belief in light of evidence—the way we should make inductive inferences—is stipulated by **Bayes' rule**. If we have hypotheses h about the state of the world, then our beliefs about those hypotheses are summarized by the probability distribution $P(h)$, known as the **prior distribution**. After observing data d, we update these beliefs to the **posterior distribution** $P(h|d)$. We can do this by applying Bayes' rule, with

$$P(h|d) = \frac{P(d|h)P(h)}{\sum_{h' \in \mathcal{H}} P(d|h')P(h')}, \tag{2.2}$$

where $\mathcal{H}$ is the set of hypotheses under consideration. The hypotheses and data vary depending on the inductive problem being solved. For categorization, the hypotheses concern the category from which an object is drawn and the features of that object are the data. For causal learning, the hypotheses are causal structures or the strength of causal relationships between variables and the data are events involving those variables.

Kahneman and Tversky had shown that people perform poorly when given probabilistic reasoning tasks that require them to use Bayes' rule (Kahneman and Tversky, 1972; Tversky, 1982).[2] But Anderson believed that rational analysis, which typically involves Bayesian inference, could nonetheless be

---

2. Gigerenzer and colleagues have argued that this can be improved by replacing probabilities with frequencies (Hoffrage and Gigerenzer, 1998; Hoffrage et al., 2015).

useful for understanding human cognition. As he put it in *The Adaptive Character of Thought*, "The basic resolution of this apparent contradiction between the results of these other fields and this book is that rationality is being used in two senses and that rationality in the adaptive sense, which is used here, is not rationality in the normative sense that is used in studies of decision making and social judgment." (Anderson, 1990, p. 31). Specifically, the applications that Anderson considered were not to explicit probabilistic reasoning tasks, but to inductive problems that people solve implicitly. His approach also took into account the structure of the environment and had room to accommodate cognitive limitations. In doing so, it allowed for the possibility that the problem that human minds are solving may not be the one that experimenters assume when comparing human behavior to the standards of classical rationality. This gap between the problem people are actually solving and the problem they are being assumed to solve makes it possible to revisit some of the evidence for human irrationality.

For example, imagine flipping a fair coin to produce a sequence of heads, Hs, and tails, Ts. Believing that getting the sequence HHTHT is more likely than getting the sequence HHHHH is clearly irrational, and is likely to lead to bad outcomes in settings such as casinos, where accurate probabilistic reasoning is important. However, there is a way to make sense of this behavior. When we ask how likely HHTHT is to be produced by a fair coin, we are asking for the probability $P(\text{HHTHT}|\text{fair coin})$. But there is another related question that people might be considering: How likely is it that this sequence was produced by a fair coin? The relevant quantity to assess to answer this second question is $P(\text{fair coin}|\text{HHTHT})$. While $P(\text{HHTHT}|\text{fair coin}) = P(\text{HHHHH}|\text{fair coin})$, it is reasonable to assume that $P(\text{fair coin}|\text{HHTHT}) > P(\text{fair coin}|\text{HHHHH})$. In fact, we can calculate this probability by applying Bayes' rule

$$P(\text{fair coin}|\text{HHHHH}) = \qquad\qquad (2.3)$$

$$\frac{P(\text{HHHHH}|\text{fair coin})P(\text{fair coin})}{P(\text{HHHHH}|\text{fair coin})P(\text{fair coin}) + P(\text{HHHHH}|\text{biased coin})P(\text{biased coin})},$$

where we assume that the coin being fair or biased are the only possibilities. Writing this out makes it clear that $P(\text{fair coin}|\text{HHTHT}) > P(\text{fair coin}|\text{HHHHH})$ because $P(\text{HHHHH}|\text{biased coin}) > P(\text{HHTHT}|\text{biased coin})$, assuming fair and biased coins are equally likely a priori. People may thus be answering this different question—one that is arguably more relevant in environments other than casinos, where being able to determine whether a sequence is random or reflects some underlying structure may be important to survival (Griffiths and Tenenbaum, 2001).

Rational analysis has been influential in cognitive science, leading to the development of Bayesian models of cognition that provide insight into a variety of inductive problems that human minds have to solve (for introductions, see Griffiths et al., 2008, 2024). However, while Anderson's rational analysis explicitly considers computational limitations in the third step of the process summarized above, it also notes that the assumed limitations should be minimal. In practice, those limitations are typically used to shape the specification of the computational problem that human minds have to solve, rather than engaging with the computational cost of the processes involved in solving those problems. This creates an opportunity: connecting the resulting rational models back to the underlying cognitive processes.

2.6 Rational process models

Understanding how abstract computational problems connect to concrete cognitive processes requires thinking about different **levels of analysis**. David Marr introduced a canonical set of three levels of analysis that apply to understanding information processing systems (Marr, 1982). The first level of analysis is the **computational level**, which focuses on the abstract computational problem the system has to solve and the ideal solution to that problem. This is the level at which rational analysis is usually conducted, with Bayesian inference being the optimal solution for a variety of inductive problems. Next, we have the **algorithmic level**, which focuses on the algorithms and representations used in executing or approximating that solution. When studying human cognition, this is the level at which cognitive psychology typically operates. Finally, there is the **implementation level**, which focuses on how those algorithms are instantiated in a physical system. For humans, this is a question of how cognitive processes are actually executed in brains.

Understanding how human minds actually solve the computational problems identified by rational analysis requires building a bridge between two levels of analysis: the computational level and the algorithmic level. Tom and his colleagues argued that one way to do so is to consider how those abstract computational problems are actually solved by computer scientists and statisticians (Sanborn et al., 2010; Griffiths et al., 2012). Given the computational costs associated with performing Bayesian inference, it is typical to use some kind of algorithm to approximate the ideal solution. These approximation algorithms provide a source of hypotheses about the cognitive processes that human minds might use. In the limit, these algorithms reproduce the rational solution, but they can be defined using known psychological processes—hence the name they gave this approach: **rational process models**.

For approximating Bayesian inference, one effective strategy is to draw samples from the posterior distribution rather than try to calculate the posterior probability of every hypothesis. Computer scientists and statisticians have developed a variety of algorithms for sampling from posterior distributions, such as importance sampling, particle filters, and Markov chain Monte Carlo (Neal, 1993; Gilks et al., 1996). These sampling algorithms can be connected to psychological processes such as those for retrieving stored exemplars from memory and stochastically searching through a space of theories, making them natural candidates for constructing rational process models.

Sampling has been used to connect rational analysis with cognitive processes in a variety of different settings. It can capture order effects in categorization and causal learning (Sanborn et al., 2010; Abbott and Griffiths, 2011), subtle phenomena of sentence comprehension (Levy et al., 2009), and errors in probabilistic reasoning (Zhu et al., 2020). It can also be used to explain how people can make what look like sophisticated Bayesian inferences based just on recalling a few pieces of relevant information from memory (Shi et al., 2010).

Rational process models indicate how we can connect the computational and algorithmic levels, linking rational analysis with cognitive processes. However, they still leave us with some important questions. In particular, even though we start with a rational model, we abandon the principle of optimization when considering how that model could be approximated. Important parameters such as the number of samples used in this approximation are typically estimated by comparing different predictions to the human data. This misses an opportunity to ask whether the approximations people use might themselves be motivated as the output of some kind of rational decision process.

2.7   Rational inattention

As rational analysis and rational process models were being developed in psychology, a parallel thread of research was exploring alternative ways to think about rationality in economics. Christopher Sims, who would go on to win a Nobel Prize for his work in macroeconomics, was puzzled why markets were less responsive to information than they should be. He proposed a model that could capture this phenomenon: if gathering information becomes increasingly costly as that information becomes more precise, rational agents could reasonably decide to limit the precision of the information they seek. The result would be decreased sensitivity to that information—a phenomenon that Sims dubbed **rational inattention** (Sims, 2003).

Rational inattention provides a framework for explaining deviations from classical rationality more generally. In particular, it can be used to account for cases where people choose to make decisions based only on partial information when that information is costly. When extended to discrete choices, this framework predicts that people's decisions will be noisy in a way that reproduces the softmax rule shown in Equation 2.1 (Matêjka and McKay, 2015).

Rational inattention illustrates how deviations from classical rationality can be captured within an optimization framework, by taking into account costs that are not reflected in the classical approach. However, the framing of rational attention in terms of the cost of information means that it is limited in the kind of psychological phenomena it can explain. Specifically, by characterizing the problem purely in terms of identifying a conditional distribution of an action given a stimulus, it still says nothing about the cognitive mechanisms that implement that mapping.

2.8   Bounded optimality and resource-rational analysis

Rational process models left us with a tantalizing question: What is the *best* way to use limited cognitive resources to approximate optimal solutions? Rational inattention suggests a path forward by formulating an alternative kind of optimization problem. But how should we express the problem of optimizing the use of our cognitive resources?

It turns out that this question was explored in the 1980s by computer scientists interested in building AI systems, most notably Stuart Russell and Eric Horvitz (Russell and Wefald, 1991a; Horvitz, 1987; Horvitz et al., 1989). These computer scientists realized that the classical theory of rational action provides a poor guide to behavior for building *any* realistic system—including those based on computers. The classical theory tells us to maximize the expected utility of our actions. But it doesn't consider the computational costs involved in calculating the expected utility associated with each action. Therefore, if an agent tried to follow the prescriptions of classical rationality, it would quickly become paralyzed as it desperately tries to compute what to do next as the world evolves around it.

Russell and Horvitz introduced a new formulation of rational action for bounded agents, which they called **bounded optimality**. The key idea is to place the criterion for rationality not on the actions an agent performs, but on the strategy the agent uses to select those actions. A bounded-optimal agent is one that uses the best algorithm to select its actions, where the best algorithm is the one that selects the highest utility actions when executed on the agent's limited computational hardware. Intuitively, attempting to compute the truly optimal action will not be bounded optimal in most cases because such an algorithm would never finish running, leading the agent to take no action at all.

Bounded optimality offers a new way to interpret deviations from classical rationality observed by psychologists and behavioral economists: these deviations may, in fact, reflect the rational use of limited cognitive resources. This is an idea that excited us. We had previously explored optimization as a way to answer questions about how many samples an agent should take (as in the work highlighting the value of a single sample mentioned in Chapter 1; Vul et al., 2009), but we began to explicitly appeal to the work of Russell and Horvitz as we began to reanalyze classic heuristics such as anchoring-and-adjustment and used it to formulate the idea of **resource-rational analysis** (Lieder et al., 2012; Chapters 3 and 4). That led to a resource-rational treatment of the availability heuristic (Lieder et al., 2014a; Chapter 4), investigation of rational metareasoning (Lieder et al., 2014b; Chapter 5), and the explicit definition of resource-rational analysis (Griffiths et al., 2015).

The same ideas were also inspiring other cognitive scientists. Richard Lewis and Andrew Howes had a long-standing interest in how constraints on cognition could guide the creation of models based on cognitive architectures (Howes et al., 2009), and formalized this idea via bounded optimality using the term **computational rationality** (Lewis et al., 2014). Thomas Icard provided a theoretical exploration of the implications of bounded optimality for studying human cognition in his dissertation (Icard, 2014a,b). Gershman et al. (2015) wrote a review article drawing on our work and related results, highlighting the potential utility of bounded optimality and related ideas for cognitive science.

2.9   Answering questions

We view resource-rational analysis as a framework that can be used to answer many of the questions left by previous approaches to thinking about the relationship between rationality and human cognition:

- Like classical rationality, resource-rational analysis is formulated in an optimization framework, but this optimization problem is expressed in terms of the internal cognitive processes of the agent and its solutions are actually achievable with finite cognitive resources.
- From bounded rationality, resource-rational analysis inherits the emphasis on cognitive constraints and the problems posed by human environments, but adds a precise mathematical definition of how a rational agent should behave subject to those constraints that can be

used to derive precise predictions in a way that has the same generality as classical treatments of rationality (see Chapter 3). This can be seen as a continuation and refinement of Baron's project to define rational thinking in a way that takes people's cognitive constraints into account. The research summarized in this book leverages modern computational methods to derive concrete rational heuristics for specific problems from a refined theory of rational thinking.

- As in the heuristics and biases framework, the expectation is that people will solve the complex computational problems posed by the environment by adopting simpler cognitive strategies—heuristics. However, resource-rational analysis provides a precise way to characterize what makes something a good heuristic: it balances the trade-off between performance and computational cost. This characterization makes it possible to understand the heuristics that people use (see Chapter 4).

- With ecological rationality, resource-rational analysis shares the idea that heuristics can be good solutions to the problems posed by human environments. It differs by emphasizing computational costs as well as the bias-variance trade-off as playing a role in selecting good heuristics, and offers a way to understand which heuristic to use when (see Chapter 5) and how to discover new heuristics (see Chapter 6).

- Like rational analysis, resource-rational analysis provides a way to derive predictions about human behavior from the solution to an optimization problem posed by the environment. However, it treats cognitive constraints as an essential part of that environment, emphasizing them to a far greater extent than rational analysis. In this way, it inherits the principle of optimization from rational analysis but pushes that principle down toward the algorithmic level, giving us a way to understand not just how people should act but how they should *think*.

- From rational process models, resource-rational analysis takes the idea that we can approximate rational solutions by assembling cognitive processes into simple algorithms. To this, it adds the idea that those algorithms might themselves be optimized, allowing us to answer the question of which algorithm offers the best way to solve a problem posed by the environment.

- Like rational inattention, resource-rational analysis provides an alternative to classical optimality that is formulated in terms of optimization. Unlike rational inattention, this optimization problem is defined over cognitive processes rather than the amount of information to seek from the environment. However, there will nonetheless be places where the two approaches overlap—as in some of the sampling-based models we discuss in this book.

Of course, resource-rational analysis raises its own set of questions. Notably: How can we actually find resource-rational solutions to meaningful problems? How well does this framework actually explain aspects of human cognition? How broad is the scope of this approach? And how can resource-rational analysis help us improve human decision-making and human lives? Our answers to these questions are presented in the remainder of the book.

3

The Rational Use of Cognitive Resources

This chapter lays out the formal framework that underpins the rest of the book: resource rationality and resource-rational analysis. As we will see, resource rationality is closely related to both bounded optimality and more traditional cost-benefit accounts of cognitive resource management. However, it is also importantly different from these ideas. We believe, and hope to convince you, that resource rationality, as presented here, provides the best formal framework for developing concrete computational models that explain how people allocate their cognitive resources, how they are able to do so much with so little, and why they sometimes fall short.

Before we get into the details, perhaps we should first address a more basic question: Why rationality? If we just want to understand and predict human cognition and behavior, why should we bother wading through a bunch of math and ideas from theoretical computer science just to figure out whether people are rational? Although there may be practical value to answering that question (see Chapter 8), we will here give an alternative motivation: rationality provides one of the most powerful tools we have for answering the most important and interesting questions about human cognition and behavior.

The benefit of using a theoretical framework, such as rationality, as a tool for cognitive modeling is that it makes decisions for us. Making a rationality assumption reduces the amount of flexibility one has when specifying a model. This may sound like an impediment—indeed, as a researcher working with rational models, it sometimes feels that way—but it can have enormous benefits. There are good reasons to think that rational cognitive models provide both a statistical edge and more satisfying explanations.

From a statistical perspective, the assumption of rationality acts as an *inductive bias*. Any given behavioral phenomenon is consistent with countless cognitive models (Anderson, 1978), but only a small subset of those represent rational solutions to problems the cognitive system could plausibly be intended to solve. If people are well adapted to their environment, then all else being equal, the rational models are more likely to resemble the truth than an arbitrary alternative model. Of course, people are not perfectly adapted to their environment; thus, one can often find a more accurate model of a particular phenomenon outside the confines of the rational framework. But when data are limited, or we are exploring a domain we know very little about, or we want to generalize our predictions in nontrivial ways, imposing a constraint that is mostly true can improve our chances of making predictions that are mostly accurate.

From a more philosophical perspective, rational models provide a type of explanation that other types of models cannot: *teleological explanations*. That is, a rational model tells us not just how a cognitive process works, but why it works that way. Characterizing a cognitive process as a rational solution to a given problem amounts to identifying that problem as the function of the cognitive process. Sometimes, that function is different from what one might have initially assumed (e.g., Anderson and Milson, 1989).

Nevertheless, as we've already seen in Chapter 2, rational models have some serious drawbacks, which parallel the two advantages we just discussed. The inductive bias of rationality is just that: a *bias*. People aren't perfectly rational, so assuming rationality means that we're going into things with an assumption that we already know to be false. On the explanatory side, while rational models are great at explaining *why*, they typically don't say as much about *how*. Classical rational models simply tell us what the rational behavior is in a given situation. They are silent about the cognitive processes that implement this stimulus-response mapping; in some sense, they're a form of behaviorism.

The question this book ultimately seeks to explore is this: Can we have our cake and eat it, too? Can we draw on the guiding principle of optimization without making the unrealistic assumption that the brain is a supercomputer? Can we explain both how the mind works and why it works that way? We propose that the answer is yes. All we need to do is think a bit more carefully about the sort of problems that cognitive systems really need to solve.

Before we begin, note that this chapter provides a narrative summary of a long history of ideas, most of which are not our own. In particular, the core idea of using tools from bounded optimality and rational metareasoning to model human cognition has been proposed and applied by many other researchers (in fact, too many to list without fear of missing one). The concept of resource rationality (and resource-rational analysis) defined here is most directly reflected in the work of Andrew Howes, Richard Lewis, and Thomas Icard—the first example of which (Howes et al., 2009) predates our earliest use of the term "resource-rational" (Lieder et al., 2012).

Additionally, in the interest of providing an exposition that is both accessible and mathematically rigorous, we will present simplified forms of some of the key concepts; we explicitly state these simplifications as they become relevant. Similarly, to maintain consistency within the chapter, we will often use different terms and notation than those used by the source material, including our own work. Again, we try to point out these discrepancies as they come up.

3.1 Formalizing rationality

What does it mean, in a formal, mathematical sense, for a cognitive system to be "rational"? Although there are many possible answers to this question, we think that the most useful way to define rationality—at least for someone trying to build or understand intelligence—is in terms of **optimality**. Unlike rationality, optimality is a well-defined mathematical notion. Something—let's call it x^*—is optimal with respect to an **objective function** f and a set of options $\mathcal{X}$ if $f(x^*)$ is the maximal possible value that f takes for any of the possible values of x. That is,[1]

$$x^* = \arg\max_{x \in \mathcal{X}} f(x). \tag{3.1}$$

3.1.1 Rational systems select rational actions: Expected utility theory

To make things concrete, let's consider a simple decision problem. You've stopped at a fruit stand on the way home from work and are deciding whether to buy an apple or a peach to have as a snack. As you may recall from Chapter 2, there's a very well-established rational model for this setting: expected utility theory. This theory states that one should take actions that yield the highest utility outcomes in expectation (roughly, on average). That is, given a **situation** s, a set of possible **actions** one could take $\mathcal{A}$, and a **utility function** U, an optimal action according to **expected utility theory** (**EUT**; Figure 3.1(a)) is defined as:

$$a_s^* = \arg\max_{a \in \mathcal{A}} \mathrm{EU}(s, a) \quad \text{where}$$
$$\mathrm{EU}(s, a) = \mathop{\mathbb{E}}_{o|s,a} [U(o)]. \tag{3.2}$$

1. Note that, for notational simplicity, we treat the argmax operator as if it returned a single item; more precisely, the argmax specifies the set of optimal items, all of which achieve the same maximal value of f. Similarly, whenever we say "the optimal X," it can be more precisely read as "the set of optimal Xs."

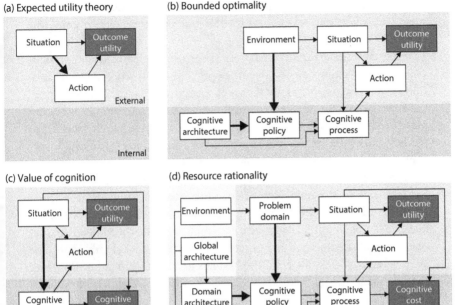

FIGURE 3.1. Four different notions of rationality for cognitive systems. Boxes and arrows indicate variables and the causal relationships between them. A bold arrow indicates that the downstream variable is optimized with respect to the upstream variable. Dark boxes indicate optimization targets. Panels (a)–(d) reflect Equations 3.2, 3.5, 3.7, and 3.10, respectively. In (d), the lightened area reflects variables that are not explicitly modeled.

Here, $\mathbb{E}_{o|s,a}$ denotes an expectation over the uncertain outcome o conditional on situation s and action a.[2] In our example, the situation s is being hungry at a fruit stand (at 5 pm, in late summer, etc.), the possible actions $\mathcal{A}$ are {apple, peach}, an outcome o is the experience of eating the fruit, and the utility function U describes the pleasure associated with that experience. Importantly, s only reflects the aspects of the situation that are knowable to the agent; thus, s includes the season but not the true flavor of any particular fruit. For this reason, there is typically uncertainty in the outcome associated with each action. However, by taking into account all the

2. If you're not familiar with this notation, you can read this expression as roughly "the average utility of the outcome, where the weights/probabilities of different outcomes are determined by the known values of s and a (see Appendix 1 for a table describing notational conventions adopted in this book).

relevant dimensions (how the peaches smell, your momentary desire for sugar, etc.), you can select the fruit that you're most likely to enjoy the most.

Importantly, note that optimality is not a property of actions themselves, or even situation-action pairs. Rather, the optimality of an action depends critically on our assumptions about the set of possible actions, {apple, peach}, and the objective function, EU(s, a). If we make different assumptions—for example, including additional options by setting $\mathcal{A} = \{$apple, peach, mango$\}$, or modeling risk aversion by setting the objective function to $\mathbb{E}_{o|s,a} U(o)^{0.9}$—we may find that a different fruit becomes the optimal choice.

Armed with the formal tool of optimality, let's return to our original problem, formalizing the rationality of cognitive systems. Although one might think that Equation 3.2 has already done just that, it actually defines an optimal *action*, not an optimal cognitive system. Thus, rather than defining a set of actions and an objective function over those actions, we should instead be defining a set of possible cognitive systems and an objective function over those systems. Given these assumptions, an **optimal cognitive system** is defined:

$$\pi^* = \arg\max_{\pi \in \Pi} f(\pi), \tag{3.3}$$

where π is one possible cognitive system, Π is the set of all possible cognitive systems, and f quantifies how well the system achieves its purpose.

Just as for actions, the optimality of a cognitive system depends critically on our assumptions about the different forms the system could have taken and the function we think it performs. Although this may seem like a limitation, these assumptions actually reflect the primary theoretical content of an optimal model. That is, when we propose an optimal model of a cognitive system, we are really making a claim about what we think the system is meant to do and what other forms it could have plausibly taken.

Returning to our fruit-stand example, can we define an optimal cognitive system for making such decisions? Taking inspiration from expected utility theory, we define our objective function as the expected utility of (the outcomes of) the actions selected by the cognitive system. For the space of possible systems, we abstract away the architectural and mechanistic details and instead characterize each possible cognitive system, π, as a stochastic mapping from situation to action and allow for any possible mapping of this form. Note that defining this set requires us to specify all possible situations and actions, both of which depend on the **environment** E that the cognitive system interacts with. We can then define the set of possible cognitive systems as $\Pi_E = \Delta(\mathcal{A}_E \mid \mathcal{S}_E)$, that is, the set of all possible conditional probability distributions over the available actions given any possible situation. Given these assumptions, the optimal cognitive system is

$$\pi_E^* = \underset{\pi \in \Delta(\mathcal{A}_E | \mathcal{S}_E)}{\arg \max} \ EU(E, \pi), \quad \text{where}$$

$$EU(E, \pi) = \underset{o|E,\pi}{\mathbb{E}} \ [U(o)] = \underset{s|E}{\mathbb{E}} \left[\underset{a \sim \pi(s)}{\mathbb{E}} \left[\underset{o|s,a}{\mathbb{E}} \ U(o) \right] \right].$$

$$(3.4)$$

That is, according to EUT, an optimal cognitive system for a given environment is one that selects actions in a way that maximizes the expected utility of the outcomes that result from those actions. We then break down that expectation into three nested expectations: first over the situation, then over the action taken in that situation, and finally over the resulting outcome.

For readers familiar with reinforcement learning, our choice of the variable π is intentional. This cognitive system is a **policy** in the technical sense: a mapping from states (situations) to actions. Note that, throughout this chapter, we will assume "one-shot" environments, in which the agent takes a single action in response to a given situation, and this action has no bearing on future situations the agent might encounter. This contrasts with a "sequential" environment in which the agent takes multiple interdependent actions. Importantly, however, we make this assumption solely to simplify the exposition. All of the ideas presented here can (and have) been applied to the sequential setting, and we address this case in the penultimate section of the chapter.

We've now defined EUT in two ways, first in terms of optimal actions and then in terms of an optimal cognitive system. So, which is it? The answer, it turns out, is both. With some squinting, you can see that the system that selects actions according to Equation 3.2 corresponds exactly to the one defined in Equation 3.4. That is, among all possible ways of making choices, always choosing actions that yield the highest expected utility given the current situation yields the maximum possible expected utility across all possible situations. The only difference between the two ways of defining EUT is what is made explicit. Equation 3.2 makes explicit how the system behaves, while Equation 3.4 makes explicit the purpose of the system, the alternative forms it could take, and the fact that the optimal cognitive system depends on the environment. In this book, we will use both notational formats to define our optimal models.

We have now defined our first optimal cognitive system, which captures a specific kind of rationality: selecting actions that result in the highest-utility outcomes (e.g., picking the tastiest fruit). This type of rationality has been called "perfect rationality" (Russell, 1997) or "Type II" rationality (Good,

1952), and is similar in spirit to consequentialist or utilitarian theories of morality (Sinnott-Armstrong, 2023; Bentham, 1961; Mill, 1998). This is certainly a desirable notion of rationality, but is it a useful one?

As a theory of rationality for cognitive systems, EUT has three major weaknesses: scientific, predictive, and computational. First, from a scientific perspective, EUT doesn't actually say anything about cognition beyond constraining possible cognitive mechanisms to those that can perform expected utility calculations. By defining a cognitive system as a mapping from situation to action, we have explicitly adopted behaviorism—ignoring the internal cognitive mechanisms that realize the behavioral mapping. Second, from a predictive perspective, we know that people don't always take perfectly rational actions (Chapter 2). Thus, assuming that people are rational in the EUT sense will lead to systematic mismatches between our models' predictions and what people actually do. Finally, from a computational perspective, EUT may not provide an appropriate target when building machines for solving the kinds of problems people encounter in the real world, which typically involve highly variable and complex situations, many possible actions, and large spaces of outcomes whose probabilities depend on the situation and the action in intricate ways. As a result, evaluating Equation 3.2 will typically require an enormous amount of computation. Thus, although one can theoretically define the rational action for any given situation, this neither tells us how people pick fruit, nor how likely they are to choose the peach, nor how to build a machine that can make this decision for us.

Fortunately, optimality is a far more general principle than EUT. In particular, EUT is a special case defined by two key assumptions about the space of possible cognitive processes (i.e., $\Pi_E = \Delta(\mathcal{A}_E \mid \mathcal{S}_E)$) and the objective function (i.e., $f = \text{EU}$). By changing one or both of these assumptions, we may be able to address the problems with EUT without sacrificing the methodological and philosophical benefits of optimality. But which assumption should we modify, and how? Putting aside important philosophical questions about whether all outcomes can be quantified and compared on a single scale, it seems to us that the assumption that one *should* select actions that maximize expected utility is on fairly solid ground. By contrast, the assumption that the human brain can implement all possible mappings from situations to actions seems much less plausible. Thus, if we want to explain deviations from expected utility theory within a rational modeling framework, perhaps we should turn a suspicious eye toward this choice of Π.

3.1.2   Accounting for cognitive constraints:
Bounded optimality

All the weaknesses of EUT discussed above have a single root cause: a fail-ure to account for the fact that people's observable behavior (e.g., Warren Buffett's investment in Goldman Sachs during the 2008 financial crisis) is gen-erated by cognitive processes that are implemented on a (bio)physical system of a limited size, whose components operate at a limited speed and are con-strained by the laws of (bio)physics and (bio)chemistry. In the language of Marr (1982), EUT is a purely computational-level theory, ignoring algorithm and implementation. How can we revise our definition of rationality to reflect the constraints imposed by physically realized computation?

Fortunately, we are not the first to ask this question. For computer sci-entists, who seek to *build* (rather than just describe) intelligent systems, implementation cannot be ignored. Although this work can proceed smoothly without worrying about esoteric notions like rationality, some researchers sought to formally define the thing that AI research was implicitly working towards. Eric Horvitz (1987) coined the term **bounded optimality**, defin-ing it as "the optimization of computational utility given a set of assumptions about expected problems and constraints in reasoning resources," and Stuart Russell refined this concept into its current version (1997; 1995). To make this concrete, we will adopt a simplified form of the definition of bounded optimality provided by Russell and Subramanian (1995). According to this definition, a **program** π is bounded optimal for a given **computational archi-tecture** C and environment E if the expected utility of the outcomes resulting from running the program on that architecture in that environment is at least as high as for any other program the architecture could have executed instead (Figure 3.1(b)).

Before giving the formal definition, let's define the terms. The environ-ment E has the same interpretation as before; it defines a distribution over possible scenarios the agent might encounter and a set of actions it can take in response. The computational architecture C corresponds to a pro-gramming language together with a machine that can interpret or implement programs in that language. Π_C is the set of all possible programs in this lan-guage, and π is one of those programs. We can also think of C as defining a set of computations the machine could execute. Adopting this view, a pro-gram π defines a situation-conditional distribution over these computations: $c \sim \pi(s)$.

Adopting the more psychological terminology of Howes et al. (2009), we can view C as a **cognitive architecture**. Each $c \in C$ is one possible **cog-nitive process** one could execute on that architecture (e.g., a sequence of

cognitive operations). π then corresponds to a "strategy" for using that architecture, determining which of the many possible cognitive processes the
person applies in any given situation. To avoid confusion with a more local
type of strategy (e.g., the strategy one uses to solve one particular problem),
we will refer to π as a **cognitive policy**. With these new concepts defined,
we can now formally define a bounded-optimal program $\pi^*_{E,C}$ as:

$$\pi^*_{E,C} = \arg\max_{\pi \in \Pi_C} \mathrm{EU}(E, C, \pi), \quad \text{where}$$

$$\mathrm{EU}(E, C, \pi) = \underset{o|E,C,\pi}{\mathbb{E}} [U(o)]$$

$$= \underset{s|E}{\mathbb{E}} \left[\underset{c \sim \pi(s)}{\mathbb{E}} \left[\underset{a|s,c}{\mathbb{E}} \left[\underset{o|s,a}{\mathbb{E}} [U(o)] \right] \right] \right].$$

(3.5)

That is, a bounded-optimal program (cognitive policy) for a given architecture
and environment is one that yields the best outcome in expectation. Similarly
to Equation 3.4 (EUT), this expectation can be broken down into four nested
expectations. The key difference is that we insert an extra step between the
situation and the action. Specifically, we account for the fact that an agent's
choice of action depends on a computational process c that results from executing program π on architecture C. Critically, unlike Equation 3.4—where
we allowed for any possible mapping from situation to action—Equation 3.5
restricts π to the set of programs that can be implemented in the architecture's language, formalized by Π_C. Keep this constraint in mind, as it will play
an important role in our story.

Comparing Equation 3.4 and Equation 3.5 also reveals a somewhat puzzling discrepancy. Equation 3.4 defines an optimal cognitive *system* π^*_E as a
mapping from situation to action, while Equation 3.5 defines an optimal cognitive *policy* $\pi^*_{E,C}$ as a mapping from situation to cognitive processes. Although
we use π for both to emphasize their shared role as an optimized source of
intelligence, there is a crucial distinction. The cognitive policy in bounded
optimality specifies how to use a cognitive architecture to guide behavior. The
cognitive system, which defines a mapping from situation to action, is defined
by an architecture-policy pair (C, π).

Compared to expected utility theory, bounded optimality shares the core
assumption that rationality ultimately grounds out in behavior and the outcomes it produces. However, in bounded optimality, that is just the beginning
(or, rather, the end) of the story. By explicitly modeling the computational processes by which behavior is produced, bounded optimality provides an answer
to the scientific, predictive, and computational weaknesses of EUT outlined
above. First (scientific), bounded optimality provides a notion of rationality

that is inherently cognitive. It is a property of how we think rather than how we act. It thus makes specific predictions about the cognitive processes that lead up to behavior. Second (predictive), bounded optimality can explain why we do not always behave perfectly rationally. Given finite computational resources, many behavioral policies are infeasible—there is no program that implements them. Typically, this includes the optimal behavioral policy; thus, a bounded-optimal cognitive system will not always produce the best action in response to every situation. Finally (computational), bounded optimality provides a rational benchmark that real cognitive systems can actually meet; in fact, it identifies the highest such benchmark. Thus, at least in theory, bounded optimality provides a useful target for both designing new intelligent systems and evaluating existing ones. For this reason, consistent with the approach taken by Lewis et al. (2014) and Gershman et al. (2015), we will adopt bounded optimality as the ideal formal definition of resource rationality.

Unfortunately, however, this is not the end of our story. Like most ideals, bounded optimality runs into major practical challenges. To see why, we must clarify what exactly the "environment" represents in bounded optimality. In psychology, people typically use this term to describe a condition of an experiment—or, if you're lucky, a type of problem that people encounter in the real world. In contrast, the environment in bounded optimality describes the entire world in which a person exists, the "situation" defines their initial position within that world (i.e., their birth), and the "action" corresponds to everything that person does in their entire life.[3] Similarly, the architecture C does not define a limited subset of cognitive activities relevant to a given task, but instead defines every conceivable thing the person could do with their brain. As a result, the program or cognitive policy π does not specify how a person directs one particular cognitive resource in the service of solving one particular type of problem, but must instead determine how they will think about any (and every) problem they might be faced with.

To illustrate why bounded optimality must adopt such overarching definitions, let's try to conduct an informal bounded-optimal analysis of our fruit-stand example. We define the environment as a distribution over possible fruit-stand problems, varying in which fruits are available, the time of year, the person's level of hunger, and so on. The bounded-optimal cognitive system for this environment is the one that selects the best fruit possible under the constraint that the system can be implemented on a human brain. This

---

3. In a more precise formulation of bounded optimality, one considers long sequences of interdependent situations and actions. More precisely, at each time step, the agent receives an observation of a latent world state, performs one step of computation, and takes one action. Equation 3.5 collapses over this temporal dimension.

is a challenging task, as any produce shopper will attest—but remember: we have an entire human brain at our disposal, and the only thing that brain will ever have to do is choose fruit. Although we cannot specify what exactly the bounded-optimal program would look like in this case, it would almost certainly carefully inspect every piece of fruit available, spending weeks or even months weighing the merits of each one. If we make the problem slightly more challenging and assume that the agent must solve millions of fruit-choice problems in its lifetime, it would certainly spend less time on each one; but it would still dedicate its entire brain to that one type of problem, doing away with any cognitive functions that are unnecessary to picking fruit.

This example is ridiculous, but we hope it drives the point home. Bounded optimality, at least in its standard form, cannot be applied to individual problems or even types of problems. This has a devastating methodological consequence. It means we cannot study tasks and cognitive systems in isolation, as almost all psychologists, economists, and neuroscientists do. Instead, we must characterize the bounded-optimal cognitive system for the human brain all in one go, which is clearly an insurmountable task. Is a bounded-optimal characterization of cognition hopeless?

3.1.3 Rational choice of mental action: VOC

To address the methodological challenges of bounded optimality, we can draw inspiration from an alternative approach to modeling cognitive constraints, one often seen in psychological models. The core idea is the concept of **cost-benefit trade-offs** (Edwards, 1954). Intuitively, thinking is effortful, and so we avoid it for the same reason we avoid carrying heavy groceries (Shenhav et al., 2017). On the other hand, more thinking generally results in better outcomes. Thus, there is a trade-off between the utility of the outcomes we experience and the cost of the mental effort we exert to attain those outcomes (Kool and Botvinick, 2018). Critically, however, the optimal trade-off is not always the same. For important decisions with clear factors to consider, thinking a lot is often worthwhile. For unimportant decisions, or ones that are too complex to reason about effectively, the benefit of thinking a lot is unlikely to outweigh the cost.

The idea that a rational cognitive system should allocate a different amount (and type) of mental effort in different situations has been formalized in several different ways (Shenhav et al., 2013; Anderson, 1990; Lieder and Griffiths, 2017; Sims, 2003). However, all these different approaches draw on the same key insight. Choosing between different possible cognitive activities that require different amounts of mental effort is analogous to choosing between products that cost different amounts of money. Thus, just as the expected utility (or "value") of buying a peach can be decomposed into the benefits of having the peach minus the monetary cost of purchasing the peach, the value of *thinking about* a peach can be decomposed into the benefits of knowing how much you want the peach minus the cognitive cost of doing that thinking.

But what are the benefits of knowing how much you want the peach? Intuitively, the value of thinking about the peach grounds out in the value of buying the peach, the expected utility of its gustatory and monetary outcomes. Thinking about the peach influences your decision to buy or forego the peach, either increasing the probability of a tangy delight or reducing the probability of a mushy disappointment (and less money in your wallet). More generally, the **benefit of cognition** derives from how it changes behavior, and ultimately the utility of the outcomes that result from this behavior. To be precise, the benefit of performing cognitive process c in situation s is the increase in the expected utility of the action we will take after thinking, relative to the utility if we acted without thinking, which we notate as $\varnothing$. Thus, we define the benefit of cognition as:

$$\Delta\mathrm{EU}(s, c) = \mathbb{E}_{a|s,c}\left[\mathbb{E}_{o|s,a}\left[U(o)\right]\right] - \mathbb{E}_{a|s,\varnothing}\left[\mathbb{E}_{o|s,a}\left[U(o)\right]\right]. \qquad (3.6)$$

Having defined the benefit of cognition, we can now ask: what is the optimal way to think about any given situation? Intuitively, one should think about that situation in the way that yields the greatest expected benefit ($\Delta\mathrm{EU}$). On the other hand, if that benefit is small, then it may not be worth paying the cost, and you would be better off not thinking at all ($\varnothing$). More generally, the value of engaging any particular cognitive process depends on both its benefit and cost. Thus, given a situation s and a set of candidate cognitive processes C the optimal cognitive process c_s^* is:

$$c_s^* = \arg\max_{c \in C} \mathrm{VOC}(s, c) \quad \text{if } \max_{c \in C} \mathrm{VOC}(s, c) \geq 0 \quad \text{and } \varnothing \quad \text{otherwise,}$$

$$\text{where } \mathrm{VOC}(s, c) = \Delta\mathrm{EU}(s, c) - \mathrm{cost}(s, c).$$
$$(3.7)$$

That is, the **value of cognition** (VOC; Figure 3.1c) is the difference between the expected increase in the utility of the action you take as a result of thinking (versus acting without thinking) and the cost of engaging in that thought. A cognitive process is optimal if it both achieves the maximal VOC (out of a set of alternative operations) and its VOC is greater than zero. In the case that no process has positive value, one should not think and simply act.

Note the parallels between Equation 3.2 (EUT) and Equation 3.7 (VOC). Both define an optimal "thing to do" in a way that depends (partially or fully) on the expected utility of the outcome it will lead to. In EUT, a "thing to do" corresponds to something we could do in the world, while in VOC, it corresponds to something we could do in our head—but there is a sense in which both of these are a form of action. Accordingly, one way to understand VOC is as an extension of EUT from *physical* action to *mental* action.

Variations of the VOC idea have been applied to model many different aspects of cognition. One influential example is the expected value of control (EVC) theory (Shenhav et al., 2013). In the EVC theory, c corresponds to a "control signal," which inhibits, initiates, or modulates a specific cognitive process. For example, it might determine the threshold or "drift rate" in a diffusion decision model (Musslick et al., 2015; Grahek et al., 2020). Importantly, this can both influence how much cognitive effort is exerted (e.g., how closely you read this chapter), and vary the nature of that cognitive effort (e.g., by focusing on the text rather than the equations). In another instantiation of the VOC idea, c corresponds to a strategy for solving a problem, for example, whether to use a "model-free" or a "model-based" decision mechanism (Daw et al., 2005). Intuitively, a model-free strategy relies on habits or learned associations, while a model-based strategy involves explicit reasoning about the consequences of one's actions. Generally, the model-based system yields better outcomes but is more costly to execute, and people seem to arbitrate between the strategies accordingly (Keramati et al., 2011; Kool et al., 2017). In Chapter 5, we present a model of strategy selection that explicitly optimizes the VOC.

The VOC idea has also been applied in artificial intelligence, specifically the field of rational metareasoning, where VOC stands for "value of computation" (Russell and Wefald, 1991b). Here c corresponds to a single "computation" that an algorithm could perform, for example, simulating a possible outcome of one available action (Hay et al., 2012). Using more psychological terminology, we could call c a "cognitive operation." Importantly, adopting this more temporally fine-grained view typically requires that multiple computations (or operations) be executed in sequence; this in turn requires our introducing the notion of a "belief state" (or mental state) that tracks the results of previous computations. We address this case in Chapter 6.

While the content and temporal scale of c varies substantially across applications of the general idea, these models all share one crucial property: the cognitive activity c indicates how one should think about a problem, but not the result of that thinking. Put another way, c encodes the *direction* of thought but not the *destination*. Thinking is useful precisely because it can change our mind in unexpected ways—if we already knew what conclusions a line of reasoning would lead to, we wouldn't have to actually follow it.

To summarize, the VOC theory suggests that a rational agent should choose how to think in a way that balances the benefits and the costs of thinking. As illustrated in Figure 3.1(c), the benefits of thinking ground out in how it influences our actions, which in turn determine our outcomes. Although we cannot know in advance precisely which action and outcome a given thought will lead to, we can estimate the benefit by taking an expectation over those uncertain actions and outcomes (hence, the "expected" in EVC). Different versions of the VOC idea underlie many of the most successful and well-known theories of how people allocate their cognitive resources.

However, we are still left with (at least) two lingering questions. First, the parallel between EUT and VOC noted above should make us somewhat uneasy. As discussed earlier, EUT sets an unrealistic normative standard because it presents an intractable optimization problem. If people cannot direct their behavior optimally, why should we expect that they can do this for their cognition? To put a finer point on this, while the VOC accounts for the cost of executing a cognitive process, it doesn't account for the cost of *selecting* that process. This is problematic because computing the expected increase in outcome utility (ΔEU) associated with a given cognitive process is generally a nontrivial problem. When we further consider that cognitive processes are not indivisible entities (as we have been assuming for simplicity) but are instead composed of long sequences of individual operations—with the results of early operations informing which operation should be executed next—the problem becomes extraordinarily complex (see Chapter 6).

The second lingering question concerns the cost term. We showed how the benefit of thought grounds out in its influence on action, but provided no such grounding for the cost. Many types of cognition certainly *feel* costly—but why? The ultimate source of cognitive cost has been a subject of much debate. Early theories appealed to explicit metabolic costs associated with neural activity (Gailliot et al., 2007), but the results supporting these theories have faced serious methodological critiques (Kurzban et al., 2013; Vadillo et al., 2016; Molden et al., 2012).

Although they seem quite different, these two questions may be deeply related. In the next section, we show how.

3.1.4 What is the cost of cognition?

As defined in Equation 3.5, bounded optimality poses a constrained optimization problem whose solution is a mind that solves all the problems the agent might face throughout its entire life in real time using a single cognitive architecture. Identifying this bounded-optimal mind thus requires finding the optimal allocation of one shared pool of cognitive resources across all possible sequences of all different types of problems one may encounter—an intimidating task, to say the least.

To get an intuition for how a bounded optimal mind might solve the problem of dynamically (re)allocating its finite cognitive resources across the many things it could think about over the course of its life, let's consider case where an organism must allocate a finite resource among several different activities: foraging. In foraging, rather than allocating mental resources across different cognitive functions, an organism must allocate its time and energy across different patches where food might be found. Initially, this problem seems very challenging because the correct amount of time to allocate to each patch depends on the time allocated to every other patch. But, as shown by Charnov (1976), the optimal solution is actually quite simple: stay in a patch as long as the rate of food intake is more than what you could expect to find elsewhere. Under the (admittedly strong) assumption that the global availability of food is constant, this quantity is simply the average rate of intake in the past. The key idea here is that one should allocate resources to an activity as long as the utility gained from that activity exceeds the **opportunity cost** of not allocating those resources to some other activity. As brilliantly observed by Kurzban et al. (2013), this principle applies equally to the allocation of cognitive resources. Thinking about any particular thing has an opportunity cost because it means you aren't thinking about (or doing) something else.

To illustrate the idea of cognitive opportunity cost, consider an extension of the fruit-stand example. Imagine that, on the way to the fruit-stand, you begin thinking about which fruit you'd like—you're leaning toward the peach. Now consider the following choice: should you think a little more about peaches, think a little more about apples, or put fruit out of your mind and think about something else? Thinking about either fruit may uncover an important nuance that you hadn't accounted for. For instance, it might help you realize that peaches really aren't the best on-the-go snack, helping you avoid a sticky situation. On the other hand, if you allow your mind to drift elsewhere, you might start thinking about something much more important, for example, how to finish the book chapter that was supposed to be done last week. Now, if we had framed the problem as an explicit choice between thinking about apples, peaches, or past-due writing projects, you would presumably have focused on the chapter. The problem is that there are countless things you could be thinking about at any moment, and you can't weigh the relative merits of all of them, not even implicitly. Something you *can* do, however, is to have a general sense of how useful thinking tends to be and redirect your mental energy when you detect that you've fallen below that baseline. Subjectively, thoughts of seasonality and ripeness might become tiresome or boring (Agrawal et al., 2022), leading you to search for greener mental pastures.

Although a full formal treatment of cognitive opportunity cost is beyond the scope of this chapter (see Kurzban et al., 2013 and Agrawal et al., 2022 for thorough examples), we provide a rough sketch of how cognitive opportunity cost can be derived from pure utility maximization under cognitive constraints. For simplicity, we here assume that each $c \in C$ corresponds to a discrete cognitive operation that consumes the agent's full cognitive capacity for the same fixed duration.

We begin by removing the cost term from the VOC (Equation 3.7), defining the optimal cognitive operation as the one that provides the most benefit (increase in expected utility; Equation 3.6):

$$c_s^* = \arg\max_{c \in C} \Delta EU(s, c).$$

Notice that this equation already captures an implicit notion of opportunity cost by virtue of the argmax operator. Executing a particular cognitive operation c has an opportunity cost associated with *not* executing any of the alternatives $c' \in C \setminus \{c\}$ at that time. Thus, assuming that C captures everything the agent could conceivably do with its brain, opportunity cost is fully accounted for. The problem is that this is a wildly unrealistic assumption because it requires that the agent (or the cognitive modeler) optimize over an enormous set of possible thoughts one could have. To avoid this intractable optimization, researchers applying the VOC framework typically work with a much smaller set of domain-specific cognitive operations C_D that are directly relevant to the **problem domain** under investigation, D. For example, in the fruit-stand domain, C_D might include "consider whether peaches are in season," but not "think of a non-self-referential example to illustrate out-of-domain operations". Unfortunately, this solution comes with its own problems. While optimizing over within-domain operations will correctly account for within-task opportunity cost, it will not account for between-task opportunity cost, that is, the possibility of thinking about something else entirely.

To address between-task opportunity cost, we define an abstract cognitive operation $\notc$ that corresponds to thinking about "something else" (e.g., a book chapter when you've been thinking about fruit). Specifically, $\notc$ is the *most beneficial* operation that is not in the set of operations we are explicitly considering:

$$\notc = \arg\max_{c \notin C_D} \Delta EU(s, c). \tag{3.8}$$

We can then derive the optimal cognitive operation by first identifying the optimal within-task operation, comparing it to the optimal out-of-task operation, and selecting the one that is better:

$$c_s^* = \arg\max_{c \in C_D} \Delta EU(s, c) \quad \text{if} \max_{c \in C_D} \Delta EU(s, c) > \Delta EU(s, \notc)$$

$$\text{and } \notc \quad \text{otherwise.}$$

Here, $\Delta EU(s, \not{c})$ corresponds to the opportunity cost of not excuting $\not{c}$. We can then move this term inside the objective function (without changing the result) to obtain:

$$c_s^* = \arg\max_{c \in \mathcal{C}_D} \Delta EU(s, c) - \Delta EU(s, \not{c}) \quad \text{if} \max_{c \in \mathcal{C}_D} \Delta EU(s, c) - \Delta EU(s, \not{c}) > 0$$

$$\text{and } \not{c} \quad \text{otherwise.}$$

Does this look familiar? Indeed, we have recovered the VOC (Equation 3.7), replacing $\varnothing$ with $\not{c}$ and defining the cost as

$$\text{cost}(s, c) = \Delta EU(s, \not{c}) = \max_{c' \notin \mathcal{C}_D} \Delta EU(s, c'). \tag{3.9}$$

That is, the cost of executing operation c in situation s is the maximal benefit one could attain by executing a different operation, one outside the set we are considering.

Having established the basic approach, we briefly address three key missing pieces. First, and most critically, we defined the "something else" operation $\not{c}$ as an argmax over the set of all out-of-task operations the agent could execute—precisely the intractable optimization we wanted to avoid. To remedy this, note that the opportunity cost depends on the benefit of performing that operation, but not on its identity; the expected benefit of this operation could be estimated in any number of ways. As in optimal foraging, simply averaging the experienced benefits of all past operations might already work quite well. In the event that $\not{c}$ turns out to be optimal, the cognitive modeler can assume that the "something else" is the next trial, leading to a prediction about response time. The situation is more complex for the agent, but one plausible option is to revert to mind wandering, searching for something else to think about or perhaps rehearsing previous thoughts. Second, recall that we defined $\Delta EU(s, c)$ under the unrealistic assumption that the agent can execute only one operation before taking a single action. However, between-task opportunity cost is felt most strongly in the sequential setting, where an agent can think about situations and actions beyond the immediate ones; a complete picture would thus require addressing this case. Finally, we made the simplifying assumption that all operations consume the brain's full capacity for the same duration. A more precise characterization would account for the fact that different operations occupy different cognitive resources for different amounts of time.

Despite these limitations, we can draw two key conclusions. First, the cognitive cost term in the VOC—and perhaps the subjective experience of mental effort—may actually reflect the foregone benefit of thinking about something else. By the same token, the choice to *stop* thinking about something ($\emptyset$ in Equation 3.7) can be understood as a decision to *start* thinking about something else ($\text{\textcent}$ in Equation 3.8). Second, cognitive cost is not an intrinsic feature of the *problem* of resource-constrained cognition—instead, it is part of the *solution* to that problem. From the bounded-optimal perspective, the agent's goal is simply to use its brain in a way that produces the best outcomes. Weighing the benefit of cognitive operations against their opportunity cost is a strategy that researchers (and perhaps animals) use to mitigate the intractability of optimizing over all the possible ways that a brain can be used.

3.1.5   *Interpolating between VOC and bounded optimality: Resource rationality*

In the previous sections, we described two quite different approaches to formalizing rationality under cognitive constraints. In the first approach (bounded optimality), rationality is a property of a program or cognitive policy that governs the behavior of an agent's cognitive architecture. According to this definition, a cognitive system is rational if it produces the highest utility outcomes of all possible systems that could be implemented on a given architecture. In the second approach (VOC), rationality is a property of a cognitive operation or a control signal. According to the VOC perspective, a cognitive system is rational if it performs operations that maximize the expected difference between outcome utility and cognitive cost. Each approach has its own strengths and weaknesses. From a philosophical perspective, bounded optimality is the clear winner. It defines exactly the problem facing a resource-bound agent. From a more pragmatic perspective, however, the cost-benefit trade-off in the VOC is much easier to work with, as evidenced by its far more widespread use in the behavioral sciences. Must we choose between principle and practice?

Perhaps not. Given that bounded optimality and VOC have complementary strengths and weaknesses, there may be a way to integrate them that balances the costs and benefits of each approach. Ultimately, all the differences between the two approaches boil down to a difference in the granularity of optimization. Bounded optimality is maximally broad, assuming that all-encompassing cognitive policies are optimized with respect to an agent's complete environment and cognitive architecture. The VOC framework is maximally narrow, assuming that individual cognitive operations (or, slightly broader, control signals) are optimized with respect to specific situations. The narrower scope of optimization means that the VOC framework is easier to apply, but it also means that we are assuming the agent itself is constantly solving those narrow (but still quite difficult) optimization problems—critically, the VOC doesn't account for the cognitive cost of solving these problems.

Characterizing VOC and bounded optimality as opposite poles on a spectrum from narrow to broad optimization suggests a clear path to finding a compromise between the two approaches: specify a notion of rationality that allows for medium-scope optimization. The previous section laid the ground for such an integration by showing how we can understand the cognitive cost term in the VOC as an approximation to between-task opportunity cost. Building on that idea, we can view a problem domain D as defining an environment of intermediate scope, smaller than the full environment but larger than a specific situation. Similarly, we can view C_D as a domain-specific cognitive architecture. However, to limit subscripts, we will use C for the domain-specific architecture and C_G for the "global" architecture. The domain-specific architecture could be a strict subset of the complete cognitive architecture, but it could also be an abstraction of that architecture, defining higher-order cognitive operations that could be broken down into simpler, more concrete operations (note that almost all applications of the VOC framework already do this). Critically, although D and C conceptually depend on their global versions (E and C_G), this relationship is implicit—we do not attempt to model E and C_G. Similarly, the opportunity cost associated with a given cognitive operation implicitly depends on the global environment and architecture.

Now, at long last, we can define a **resource-rational** cognitive policy for a problem domain D (embedded in environment E) and domain-specific cognitive architecture C (embedded in global architecture C_G) as:

$$
\begin{aligned}
\pi_{D,C}^* &= \arg\max_{\pi \in \Pi_C} \mathrm{RR}(\pi, C, D), \quad \text{where} \\
\mathrm{RR}(\pi, C, D) &= \mathop{\mathbb{E}}_{s|D}\left[\mathop{\mathbb{E}}_{c \sim \pi(s)} \left[\mathop{\mathbb{E}}_{a|s,c} \left[\mathop{\mathbb{E}}_{o|s,a} [U(o)] \right] - \mathrm{cost}(s,c) \right] \right].
\end{aligned}
\tag{3.10}
$$

That is, the resource-rational cognitive policy for a given domain and architecture, $\pi_{D,C}^*$, is the one that achieves the optimal trade-off between the expected utility of the actions it takes and the opportunity cost of the cognitive processes that produce those actions—all under the constraints imposed by the architecture (Figure 3.1(d)). The resource-rational cognitive system is then defined by the architecture-policy pair $(\pi_{D,C}^*, C)$.

Resource rationality represents an interpolation between VOC and bounded optimality that allows the cognitive modeler to achieve a good balance between the tractability of the VOC framework and the accuracy of the bounded optimality framework. Resource rationality reduces to bounded optimality when $D = E$ and $C = C_G$. In this case, the opportunity cost goes to zero because there are no out-of-domain operations (or processes). Conversely, resource rationality reduces to the VOC framework when at least one of two conditions is met. First, if the domain comprises only a single situation ($D = \{s\}$), then the resource-rational cognitive policy specifies one specific cognitive process: c_s^*, as defined in Equation 3.7. Second, if the cognitive architecture can implement all possible mappings from situations to cognitive processes ($\Pi_C = \Delta(C \mid S)$), then the resource-rational policy will be the mapping $s \mapsto c_s^*$ (more on this soon). In both cases, the policy always executes the cognitive process that is optimal according to VOC.

We can also view a resource-rational cognitive system as an approximation to the component of the full bounded-optimal system that is directly relevant to the specified domain. It is approximate for three reasons. First, it rests on the assumption that an agent's global cognitive architecture can be meaningfully divided into distinct components (and that we have correctly identified and modeled one such component). Second, even assuming a perfect partitioning, the cognitive cost term may not fully account for interactions between the domain of study and other problem domains. Finally, the set of cognitive policies Π_C may not correctly capture the set of cognitive policies that an agent could realistically adopt. The first two sources of approximation error are shared by all tractable notions of rationality for cognitive systems. However, the third is specific to resource rationality (and bounded optimality), and thus merits further discussion.

The problem of specifying Π_C is one of identifying a level of expressivity that balances two competing desiderata. On the one hand, we want to capture the flexible and adaptive cognitive processes that we observe in humans. On the other hand, we do not want to allow for unrealistic degrees of adaptation. To make this latter concern concrete, consider again the case where $\Pi_C = \Delta(C \mid S)$. In this case, the policy can be any mapping from situation to cognitive process. The VOC defines the optimal such mapping, and so the policy will simply maximize VOC. However, as previously noted, maximizing VOC is generally more difficult than simply maximizing expected utility; it is an entirely unrealistic model of how people work around cognitive constraints. To address this, researchers (in both computer science and cognitive science) typically specify a much narrower space of possible programs. For example, Howes et al. (2009) specify a small set of programs corresponding to interpretable strategies in an ACT-R architecture. However, this approach can easily err in the opposite direction, failing to capture the full spectrum of mental strategies that people could use. An alternative approach is to specify a more generic constraint on the space of cognitive policies. For example, we can assume that the cognitive policy must be represented by a program with a limited number of instructions. A third approach, which we see as especially promising, is to assume that cognitive policies are learned through experience. Not only does this provide a relatively agnostic way to limit the flexibility of cognitive policies, it also begins to address a higher order question of how people might actually approach a resource-rational benchmark.

Taking a step back, resource rationality differs from expected utility theory, bounded optimality, and the VOC framework in several key ways (Figure 3.1). While expected utility theory (Figure 3.1(a)) defines the optimal *behavior* of idealized agents with infinite computational resources, resource rationality defines optimal *cognitive systems* for real people with limited time and bounded cognitive resources. Thus, unlike expected utility theory, resource rationality can explain why people don't always make optimal decisions—especially in situations where the stakes are low and the opportunity cost is high, such as when you are passing a fruit stand on your way to an important meeting that you still have to prepare for.

In contrast to bounded optimality (Figure 3.1(b)), which defines a global cognitive system that maximizes a bounded agent's success across all the situations it might encounter in its environment, resource rationality defines a specific cognitive system that can closely approximate the cognitive processes a bounded-optimal agent would apply to a particular type of problem, such as deciding which fruit to buy. This decomposition is made possible by accounting for the opportunity cost associated with directing mental resources to the current problem rather than any of the other problems in the agent's broader environment. Thus, although resource rationality and bounded optimality can both explain why people don't always make optimal choices *in principle*, resource rationality can be much more readily applied to understand a cognitive system as complex as the human mind *in practice*.

Comparing resource rationality with the VOC framework (Figure 3.1(c)) reveals subtler—but crucial—differences. Within the cognitive science literature, models that trade off utility and cognitive cost (as prescribed by VOC) are the closest theoretical neighbors to the resource-rational models that we present in this book. However, while VOC is a property of individual cognitive processes considered *in isolation*, resource rationality is a property of the cognitive policies that select or guide those cognitive processes; it reflects how well the cognitive system performs *in general*, across all the different problems it might face. As a result, unlike VOC, resource rationality does not postulate that people always execute the cognitive process that achieves the best trade-off between outcome utility and cognitive cost.

3.1.6 The sequential case

Recall that, throughout this chapter, we have focused on the "one-shot" case, in which an agent takes a single action after executing a unitary cognitive process (that is, one whose internal structure is predetermined and not subject to the optimization defined in Equation 3.10). Most of the applications in this book have this structure (except for some applications in Chapter 6 and Chapter 8). A complete theory of resource rationality, however, must account for the fact that both cognition and behavior occur over time. We refer to this problem of choosing a sequence of cognitive operations/processes and/or actions as "the sequential case."

To extend resource rationality (Equation 3.10) to the sequential case, we must account for the fact that both the true nature of a situation and the agent's understanding of that situation can change over time. Both of these processes—physical and mental—are only partially under the agent's control. In classic formulations of this problem (e.g., Kaelbling et al., 1998), both processes are shaped by the agent's physical actions: opening a box allows us to both collect and observe its contents. In a resource-rational formulation, we additionally consider the agent's *mental* actions (cognitive operations): we may recall that a venomous snake was recently placed in the box and decide not to open it after all.

To capture these dynamics, we begin by replacing the singular action and cognitive process with sequences of actions $(a_1, a_2, \ldots, a_t, \ldots)$ and cognitive operations $(c_1, c_2, \ldots, c_t, \cdots)$. We then introduce two new dynamic variables pertaining to the external world and the agent's internal mental state, respectively. The **world state** w_t captures the true state of the external world (as relevant to the current situation); its dynamics are shaped by the agent's external actions $(p_D(w_t \mid w_{t-1}, a_{t-1}))$. The **mental state** m_t captures the agent's current representation of the situation; its dynamics are shaped by the agent's cognitive operations $(p_C(m_t \mid m_{t-1}, c_{t-1}, o_t))$. Note that these dynamics also depend on the **observation** o_t which captures the agent's (incomplete) perception of the world state. At each time step, the agent selects both an action and a cognitive operation based on its current representation of the problem: $(a_t, c_t) \sim \pi(m_t)$. Importantly, however, the action can be to "do nothing," allowing the agent more time to deliberate. Finally, the outcome is simply the final world state w_N when the situation is resolved (at time step N); this outcome has an associated utility, $U(w_N)$. We can, therefore, generalize our definition of resource rationality to the sequential case as follows:

$$
RR(\pi, \mathcal{C}, D) = \underset{\substack{(a_t, c_t) \sim \pi(m_t) \\ m_t \sim p_C(\cdot \mid m_{t-1}, c_{t-1}, o_t) \\ (w_t, o_t) \sim p_D(\cdot \mid w_{t-1}, a_{t-1})}}{\mathbb{E}} \left[U(w_N) - \sum_{t=1}^{N} \text{cost}(m_t, c_t) \right].
$$

(3.11)

Although this equation bears little superficial resemblance to Equation 3.10, it is really just a slightly more detailed statement of the same idea. In both cases, resource rationality is defined by a trade-off between the utility of the outcomes a cognitive system attains $(U(w_N)$ versus $U(o))$ and the cost of the cognitive process underlying the behavior that produced those outcomes (the summation versus the single cost term). Chapter 6 examines the sequential case in detail. Here, we will only highlight two key features of the sequential case that are particularly relevant to the preceding discussion.

First, in the sequential case, the "cost" of cognition includes the risk that certain opportunities may vanish if you don't seize them promptly enough as well as the familiar between-situation opportunity cost. In the sequential model, time moves forward at a constant rate—we cannot "pause" to think. Of course, we can sit still and think about a problem, but the situation may change dramatically while we deliberate (for example, the fruit stand could close). These two types of cost are functionally very different. Within-situation costs are highly non-linear but have a locally defined scope. Between-situation costs have a very broad scope (perhaps, one's entire life) but can often be approximated by a linear function of the amount of deliberation.[4] Crucially, the cost terms in Equations 3.10 and 3.11 only capture the between-situation opportunity costs. Within-situation costs are implicitly embedded in $p(a|s, c)$ in Equation 3.10 (if the time of action is included in a) and explicitly in p_D in Equation 3.11 (through changes in the world state when the "do nothing" action is taken).

Second, considering the sequential nature of real-world situations raises a key question: over what timescale should a single "situation" be defined? Put another way, at what point can we say that one situation has ended and a new one has begun? Formally, a situation corresponds to a problem that can be considered independently of other problems.[5] In reality, no situation a person

---

4. This is because the nonlinearities of the effects on other situations average out due to the sheer number of those situations; cf. Houston and McNamara (1999).

5. This is similar to the concept of an "episode" in reinforcement learning (Sutton and Barto, 2018).

faces is entirely independent of all others, at least not if we limit ourselves to a tractable definition of the world state. However, some boundary lines will yield better approximations than others; identifying those boundaries is a primary task for the cognitive modeler (and perhaps also for the agent itself; cf. Zacks and Tversky, 2001).

These two points—concerning cognitive cost and the challenge of interdependence—are the sequential equivalent of the key ideas that motivated resource rationality as an approximation to bounded optimality. And it is the relationship between these two ideas that gives resource rationality its name.

Understanding cognition as a *resource*, like time or money, makes the challenge of interdependence enormously less challenging. This is because a limited resource entails a particular kind of global constraint, one that can often be well approximated by a local cost. Limited resources impose global constraints because they must be allocated across many different areas (global) and one simply cannot allocate more than one has (constraints). In some situations, such as day planning and budget meetings, we do actually reason about time and money in this way—as constraints. But if we did this all the time, considering how each purchase impinges on every other possible purchase, both the field of economics and economic choice itself would be utterly intractable. Treating a constraint as a cost breaks these complex interdependencies, allowing us to decompose the impossibly complex problem of allocating all of one's money into many small cost-benefit trade-offs. At its core, resource rationality is simply the application of this idea to the problem of making good use of one's brain.[6]

3.1.7 Summary

To summarize the chapter so far, we formalized the rationality of cognitive systems in terms of optimality: a rational cognitive system achieves the maximal value of an objective function out of a set of possible systems. We then introduced bounded optimality (Equation 3.5), which constrains the set of possible cognitive systems to those that can be implemented in the organism's brain. We saw that identifying the bounded-optimal cognitive system for human cognition is intractable because it requires jointly optimizing for all

---

6. In this sense, resource rationality could be viewed as an application of Lagrange multipliers (which convert hard constraints into smooth costs) to the constrained optimization problem posed by bounded optimality. Lagrange multipliers are also important for information-theoretic models of resource-constrained cognition (Ortega and Braun, 2013; Matějka and McKay, 2015; Sims, 2016).

problems the mind has to solve. However, this intractably large optimization problem can be decomposed into a set of smaller optimization problems for specific domains. The bounded-optimal mind can then be approximated by determining the optimal trade-off between the quality of the solutions to the smaller problems and the opportunity cost of the foregone benefit of applying one's cognitive resources to another domain or situation (Equation 3.10). The question we turn to now is: how can we apply what we have learned to develop better rational models of cognition?

3.2 Resource-rational analysis

Resource-rational analysis is a five-step methodology that leverages the formalization of rational cognitive processes introduced above to derive process models of cognitive abilities from formal definitions of their function and assumptions about the mind's cognitive architecture. As the name suggests, our approach is based on John Anderson's (1990) rational analysis. Rational analysis is defined by four key steps: (1) specify the problem the cognitive system is intended to solve, (2) derive the optimal solution to that problem, (3) compare the model to empirical data, and (4) refine the assumptions made in step 1 to resolve discrepancies. A key principle of this approach is to make "minimal" assumptions about cognitive limitations. That is, rational analysis is based on the working assumption that cognitive systems approach perfect rationality (unbounded optimality). In our extension of this approach, we instead make quite substantial assumptions about cognitive limitations. Formally, we assume that cognitive systems approach bounded optimality (which we practically approximate as resource rationality). In practice, this amounts to inserting an extra step (step 2) into the four steps specified above and taking the cost of cognition into account. The five steps of resource-rational analysis are thus:

1. Formalize a problem domain that a cognitive system is optimized for as a distribution over situations the agent might encounter $P(S)$, a set of actions the agent can take in response $\mathcal{A}$, and a utility function U.

2. Posit a class of cognitive systems that the brain could implement to solve this type of problem as a set of domain-specific cognitive operations $\mathcal{C}$, the opportunity cost of those operations, and a space of feasible cognitive policies for selecting among them $\Pi_{\mathcal{C}}$.

3. Identify the cognitive system in this class that maximizes expected utility minus expected opportunity cost (Equation 3.10).

4. Derive behavioral predictions from the model and compare those predictions to human behavior.

5. Address discrepancies by revising the problem definition (step 1) or possible cognitive system (step 2); alternatively, if the assumptions are already sufficiently realistic, stop.

Naturally, our approach has much in common with Anderson's original approach. In particular, it begins with a computational-level analysis (Marr, 1982) that describes the function of the cognitive system: the problem it is designed to solve. However, while the remaining steps share the same basic structure as rational analysis, putting computational limitations at the forefront raises new challenges and opportunities.

The key new theoretical challenge introduced by resource-rational analysis is the specification of the space of possible cognitive systems (step 2). Our general strategy here is based on specifying a domain-specific architecture, which we have previously called an abstract computational architecture (Griffiths et al., 2015). This architecture is defined by a set of elementary operations that the mind might use to solve the problem, along with their costs (cf. Simon, 1979; Posner and McLeod, 1982; Chase, 1978). Importantly, in contrast to cognitive architectures like ACT-R (Anderson, 1996) and SOAR (Laird et al., 1987), we specifically do *not* try to identify low-level primitive operations that could be applicable in any domain of cognition—at least, not at first. Instead, we attempt to identify a small set of more abstract operations that can nevertheless capture a range of cognitive processes. In most cases, we design the space of operations such that perfectly rational behavior can be produced in the limit of executing infinitely many operations. Thus, when we set the opportunity cost to zero, resource-rational models often reduce to a classically rational model.

For example, in Chapters 4 and 6, we adopt cognitive architectures based on one core operation: drawing a noisy sample of the subjective value of a choice option. This architecture abstracts away from the question of how such samples might actually be produced (perhaps through memory recall or mental simulation). This simplification is intentional. By starting with the simplest possible architecture, we can establish a baseline level of empirical fit, which allows us to make informed decisions about how to make the architecture more realistic in further iterations.

The next key challenge is a technical one concerning the identification of an optimal cognitive process in step 3. The optimization step often presents a challenge in classical rational analyses; cognition is full of intractable problems. But, perhaps counterintuitively, this problem is even more severe for resource-rational analysis. Although the *solution* to the bounded optimization problem may demand minimal cognitive resources, the problem of *finding* that solution is extraordinarily difficult.

Another challenge—which is really more of an opportunity—lies in step 4, where we compare model predictions to human behavior. As we mentioned above, classical rational models are typically defined as input-output mappings. Such models are thus typically limited to predicting people's overt behavior, for example, their choices. In contrast, the models derived by resource-rational analyses predict not just choices, but also the full sequences of cognitive operations leading to those choices. Thus, at a minimum, we can evaluate models in terms of their ability to predict response times. However, in cases where there is more than one possible cognitive operation, response times alone are a highly sparse signal. This motivates the use and development of *process-tracing* paradigms, which provide a more complete readout of the cognitive operations a person actually executed before making a decision. Process-tracing paradigms will play a key role in Chapter 6.

The final challenge is determining when to stop refining the assumptions about cognitive constraints. The purpose of revising the assumptions about cognitive constraints and the problem to be solved is to make them more accurate. However, we do not expect that people will be perfectly resource rational. Therefore, perpetually tweaking those assumptions until the model fits the data perfectly would likely lead to overfitting, that is, introducing incorrect assumptions about people's cognitive constraints. That's the opposite of what this step is meant to achieve. To avoid this failure mode, it is important to stop iterating once all known or theoretically plausible cognitive constraints have been incorporated.

In the next chapter, we illustrate how we have applied this framework to develop models of the cognitive processes underlying judgment and decision-making across various domains.

4

(Ir)rationality Revisited

Resource-rational analysis starts from a different set of assumptions than classical rationality. This allows us to revisit some of the ways in which people systematically deviate from classical rationality, and ask whether we can understand them from a resource-rational perspective. This chapter demonstrates that this is indeed the case, suggesting that resource rationality may be better suited to understand human cognition and predict human behavior than classical rationality.

The assumptions behind resource-rational analysis differ from those of classical rationality in three major ways. First, reasoning is evaluated by its utility for subsequent decisions rather than by its formal correctness—a kind of pragmatism not captured in classical notions of rational belief updating, such as logic and probability theory. Second, it takes into account the cost of time and the boundedness of people's cognitive resources. Third, rational actions are defined with respect to the distribution of problems in the environment rather than a set of arbitrary laboratory tasks. Arguably, all three changes are necessary to obtain a normative yet realistic standard for human behavior. Unlike in classical rationality, choices are not evaluated by the quality of people's actions or the truthfulness or coherence of their beliefs; rather, they are evaluated in terms of the underlying cognitive strategies. The quality of these strategies is not assessed by their adherence to rules that preserve truth or coherence, but, rather, by their practical impact on people's actions and their consequences.

Adopting this perspective provides a new way of thinking about the deviations from classical rationality observed by Kahneman and Tversky. Observing that human behavior is inconsistent with classical rationality doesn't necessarily imply that people don't make intelligent choices; rather, it might tell us that the standard we have been using to evaluate those choices is incorrect. As Kahneman himself put it, "Irrational is a strong word, which connotes impulsivity, emotionality, and a stubborn resistance to reasonable argument.

I often cringe when my work with Amos is credited with demonstrating that human choices are irrational, when in fact our research only showed that Humans are not well described by the rational-agent model" (Kahneman, 2011, p. 411).

In particular, the idea that people might use heuristics to overcome some of the computational costs involved in classical rationality is very well aligned with the ideas behind resource-rational analysis. In fact, resource-rational analysis can be viewed as defining what makes something a good heuristic. To be a good heuristic, a cognitive strategy has to strike the right balance between achieving good outcomes and managing computational cost. Revisiting some of these classic heuristics with the tools of resource-rational analysis thus offers a new perspective on why people use them.

We chose to begin this process by considering two heuristics whose discovery set fire to the picture of man as a rational animal: anchoring-and-adjustment and availability. These heuristics give rise to the anchoring bias and availability biases in judgment and decision-making, respectively. These biases produce violations of fundamental axioms of rationality, including the violation of the axiom of independence of irrelevant alternatives demonstrated by the Allais paradox. By examining whether these heuristics—and the resulting biases—can be reconciled with resource rationality, we have a strong opportunity to explore the potential of this approach to offer a new way of thinking about human cognition.

4.1 Case study 1: A resource-rational perspective on the anchoring bias

Our first case study highlights the four steps of resource-rational analysis from Chapter 3 in a study that sought to uncover the computational mechanisms that give rise to the anchoring bias introduced in Chapter 2.

Anchoring-and-adjustment is a heuristic people use to estimate numerical quantities. Therefore, the first step in our resource-rational analysis of anchoring-and-adjustment (Lieder et al., 2018b) was to define the problem solved by numerical estimation. The second step was to posit which kind of computational architecture the mind might employ to solve this problem. The third step was to derive the optimal solution to the numerical estimation problem afforded by the computational architecture. Finally, the fourth step was to evaluate the resulting predictions against people's estimates of numerical quantities under various experimental conditions.

4.1.1 Step 1: Formalize the problem domain that a cognitive system is optimized for

In numerical estimation, people have to make an informed guess about an unknown quantity X based on their knowledge K. In general, people's relevant knowledge K is incomplete and insufficient to determine the quantity X with certainty. For instance, people asked to estimate the boiling point of water on Mount Everest typically do not know its exact value, but they do know related information, such as the boiling point of water at normal altitude, the freezing point of water, the qualitative relationship between altitude, air pressure, and boiling point, and so on. To express the problem of numerical estimation in the language of Chapter 3, we can say that the action is to state an estimate $\hat{x}$, and the utility of that action depends on the true value of x in the current situation s. Concretely, we assume that the utility of stating an estimate decreases as the difference between x and $\hat{x}$ increases. That is, the larger the error, the higher the cost of the error and the lower the resulting utility.

We can also specify the problem at a slightly higher level of abstraction. To do so, we formalize people's uncertain belief about X by the probability distribution, $P(X \mid K)$, which assigns a plausibility $p(X = x \mid K)$, to each potential value x. According to Bayesian decision theory, the goal is to report the estimate $\hat{x}$ with the highest expected utility:

$$U(K, \hat{x}) = \mathbb{E}_{x|K}[-\text{cost}(\hat{x}, x)]. \tag{4.1}$$

Here, K corresponds to the "situation" of knowing certain things but not others, and $\hat{x}$ corresponds to an "action." The optimal (perfectly rational) estimate is thus defined by

$$x_K^\star = \arg\min_{\hat{x}} \mathbb{E}_{x|K}[\text{cost}(\hat{x}, x)]. \tag{4.2}$$

4.1.2 Step 2: Posit a class of cognitive policies that the brain could implement to solve this type of problem, along with the cost of the cognitive resources used by those systems

How the mind should solve the problem of numerical estimation (see Equation 4.2) depends on its computational architecture, $\mathcal{C}$. Thus, to derive predictions from the assumption of resource rationality, one has to specify the mind's elementary operations and their cost. To do so, we modeled the cognitive resources available for numerical estimation within a formal computational framework that has been successfully used to develop rational process models of human cognition and can capture the variability of human performance, namely, **sampling** (Griffiths et al., 2012). Sampling is widely used to solve infer-ence problems in statistics, machine learning, and artificial intelligence (Gilks et al., 1996). It is especially suitable for problems involving uncertainty. Many different sampling algorithms have been developed to solve a wide range of different problems, including those for estimating unknown values, inferring the causes of noisy observations (perception and causal reasoning), general-izing, reasoning under uncertainty, making predictions, decision-making, and learning models and theories about how things work (Griffiths et al., 2024).

All sampling methods rely on the same fundamental operation: stochas-tically simulating the outcome of an event or the value of a quantity such that, on average, the relative frequency with which each value occurs corre-sponds to the probability distribution that the algorithm is sampling from. Several behavioral and neuroscientific experiments suggest that the brain uses computational mechanisms similar to sampling for a wide range of inference problems, ranging from vision to causal learning (e.g., Vul et al., 2014; Deni-son et al., 2013; Bonawitz et al., 2014b,a; Griffiths and Tenenbaum, 2006; Stewart et al., 2006; Fiser et al., 2010; Gong and Bramley, 2023). One piece of evidence is that people's estimates of everyday events are highly variable even though the average of their predictions tends to be very close to the optimal estimate prescribed by Bayesian decision theory (see Equation 4.2, Griffiths & Tenenbaum, 2006; 2011). Moreover, both the neural variability (Fiser et al., 2010) and the behavioral variability (Vul et al., 2014; Denison et al., 2013; Griffiths and Tenenbaum, 2006) that human and non-human ani-mals show in the face of uncertainty is strikingly similar to the variability across the different samples generated by a sampling algorithm. For instance, people's predictions of an uncertain quantity X given partial information y are roughly distributed according to its posterior distribution $p(X \mid y)$, as if they were sam-pled from it (Griffiths and Tenenbaum, 2006; Vul et al., 2014). Such variability has also been observed in decision-making: in repeated binary choices from experience, animals chose each option stochastically with a frequency roughly proportional to the probability that it will be rewarded (Herrnstein and Love-land, 1975). This pattern of choice variability, called **probability matching**, is consistent with the hypothesis that animals perform a single simulation and choose the simulated action whenever its simulated outcome is posi-tive. People also exhibit probability matching when the stakes are low, but as the stakes increase, their choices transition from probability matching to maximization (Vulkan, 2000). This transition might arise from people gradu-ally increasing the number of samples they generate to maximize the amount of reward they receive per unit time (Vul et al., 2014). In summary, sam-pling can solve most problems posed by the uncertainties people encounter in everyday life, and the variability of people's responses is often reminiscent of sampling. We, therefore, assume that sampling is one of the fundamental cognitive operations of the human mind.

According to Vul et al. (2014), people may estimate the value of an unknown quantity X using only a single sample from the subjective probability distribution $P(X \mid K)$ that expresses their beliefs. If the expected error cost (Eq. 4.2) is approximated using a single sample $\tilde{x}$, then that sample becomes the optimal estimate. Thus, the observation that people report estimates with frequency proportional to their probability is consistent with their approximating the optimal estimate using only a single sample.

However, for the complex inference problems that people face in everyday life, generating even a single perfect sample can be computationally intractable. Thus, while sampling is a first step from computational-level theories based on probabilistic inference toward cognitive mechanisms, a more detailed process model is needed to explain how simple cognitive mechanisms can solve the complex inference problems of everyday cognition. We explored a more fine-grained model of mental computation whose elementary operations serve to approximate sampling from the posterior distribution (Lieder et al., 2018b). In statistics, machine learning, and artificial intelligence, sampling is often approximated by **Markov chain Monte Carlo (MCMC)** methods (Gilks et al., 1996). MCMC algorithms allow the drawing of samples from arbitrarily complex distributions using a stochastic sequence of approximate samples, each of which depends only on the previous one. Such stochastic sequences are called Markov chains; hence the name Markov chain Monte Carlo.

We assumed that the mind's computational architecture supports MCMC by two basic operations. The first operation takes in the current estimate and stochastically modifies it to generate a new one. The second operation compares the posterior probability of the new estimate to that of the old one and accepts or rejects the modification stochastically. Together, these two basic operations are one step (iteration) of an effective MCMC strategy for probabilistic inference known as the Metropolis-Hastings algorithm (Hastings, 1970). This algorithm is the basis for our anchoring-and-adjustment models, as illustrated in Figure 4.1. In other words, we assume that the class of cognitive processes the brain can implement to compute numerical estimates (Π) is equivalent to a set of Metropolis-Hastings algorithms that differ primarily in the number of iterations they perform.

We further assumed that the cost of cognition is proportional to how many such operations have been performed. Because the two elementary operations are always performed together, this is equivalent to assuming that each iteration c in which a potential adjustment is generated, considered, and potentially performed always incurs the same amount of cognitive cost.

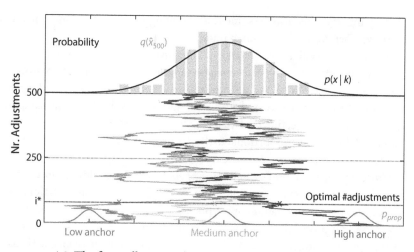

FIGURE 4.1. The figure illustrates the resource-rational anchoring-and-adjustment. The three jagged lines are examples of stochastic sequences of estimates the adjustment process might generate starting from a low, medium, and high anchor, respectively. In each iteration, a potential adjustment is sampled from a proposal distribution p_{prop} illustrated by the bell curves. Each proposed adjustment is stochastically accepted or rejected such that over time the relative frequency with which different estimates are considered, $q(\hat{x}_t)$, becomes the target distribution $p(x \mid k)$. The top of the figure compares the empirical distribution of the samples collected over the second half of the adjustments with the target distribution $p(x \mid k)$. Importantly, this distribution is the same for each of the three sequences. In fact, it is independent of the anchor, because the influence of the anchor vanishes as the number of adjustments increases. Yet, when the number of adjustments (iterations) is low (e.g., 25), the estimates are still biased towards their initial values. The optimal number of iterations i^* is very low, as illustrated by the dotted line. Consequently, the resulting estimates indicated by the X marks are still biased towards their respective anchors.

To be concrete, given an initial guess $\hat{x}_0$, which can be interpreted as the anchor a ($\hat{x}_0 = a$), this algorithm performs a series of adjustments. In each step, a potential adjustment δ is proposed by sampling from a probability distribution. The adjustment will either be accepted, that is, $\hat{x}_{t+1} = \hat{x}_t + \delta$, or rejected, that is, $\hat{x}_{t+1} = \hat{x}_t$. If a proposed adjustment makes the estimate more probable, then it will always be accepted. Otherwise, the adjustment will be made with probability $\alpha = \frac{P(X = \hat{x}_t + \delta \mid K)}{P(X = \hat{x}_t \mid K)}$, that is, according to the posterior probability of the adjusted relative to the unadjusted estimate. This strategy ensures that regardless of which initial value you start from, the frequency with which each value x has been considered

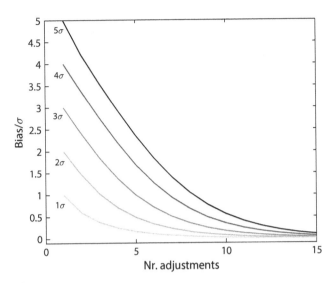

FIGURE 4.2. In resource-rational anchoring-and-adjustment the bias of the estimate is bounded by a geometrically decaying function of the number of adjustments. The plot shows the bias of resource-rational anchoring-and-adjustment as a function of the number of adjustments for five different initial values located $1, \ldots, 5$ posterior standard deviations (i.e., σ) away from the posterior mean. The standard normal distribution was used as both the posterior $P(X \mid K)$ and the proposal distribution $P_{\text{prop}}(\delta)$.

will eventually equal its subjective probability of being correct, that is, $P(X = x \mid K)$. This is necessary to capture the finding that the distribution of people's estimates is very similar to the posterior distribution $P(X = x \mid K)$ (Vul et al., 2014; Griffiths and Tenenbaum, 2006). More formally, as the number of adjustments t increases, the distribution of estimates $Q(\hat{x}_t)$ converges to the posterior distribution $P(X \mid K)$. This model of computation has the property that each adjustment decreases an upper bound on the expected error by a constant multiple (Mengersen and Tweedie, 1996). This property is known as geometric convergence and is illustrated in Figure 4.2.

There are several good reasons to consider this computational architecture as a model of mental computation in the domain of numerical estimation. First, the success of MCMC methods in statistics, machine learning, and artificial intelligence suggests they are well suited for the complex inference problems people face in everyday life. Second, MCMC can explain important aspects of cognitive phenomena ranging from category learning (Sanborn et al., 2010) to the temporal dynamics of multistable perception

(Moreno-Bote et al., 2011; Gershman et al., 2012), causal learning (Bramley et al., 2017), causal reasoning in children (Bonawitz et al., 2014a), hypothesis generation (Dasgupta et al., 2017) and revision (Fränken et al., 2022; Bramley et al., 2017), learning mental models (Ullman et al., 2018), and developmental changes in cognition (Bonawitz et al., 2014b). Third, MCMC is biologically plausible in that it can be efficiently implemented in recurrent networks of biologically plausible spiking neurons (Buesing et al., 2011). Last but not least, process models based on MCMC might be able to explain why people's estimates are both highly variable (Vul et al., 2014) and systematically biased (Tversky and Kahneman, 1974).

4.1.3 Step 3: Identify the cognitive policy that maximizes expected utility minus expected opportunity cost

To derive the resource-rational heuristic for numerical estimation, we made three assumptions. First, the estimation process is a sequence of adjustments such that, after sufficiently many steps, the estimate will be a representative sample from the belief $p(x \mid k)$ about the unknown quantity x given the knowledge k. Second, each adjustment costs a fixed amount of time. Third, the number of adjustments is chosen to achieve an optimal speed-accuracy trade-off. It follows that people should perform the optimal number of adjustments, that is,

$$t^{\star} = \arg\min_{t} \left[\mathbb{E}_{\hat{x}_t} \left[\mathrm{cost}(x, \hat{x}_t) + \gamma \cdot t \right] \right], \tag{4.3}$$

where x is its unknown true value, $\hat{x}_t$ is the estimate after performing t adjustments, $\mathrm{cost}(x, \hat{x}_t)$ is its error cost, and γ is the time cost per adjustment.

Figure 4.3 illustrates this equation, showing how the expected error cost—which decays geometrically with the number of adjustments—and the time cost—which increases linearly—determine the optimal speed-accuracy trade-off. We inspected the solution to Equation 4.3 when the belief and the proposal distribution are standard normal distributions (i.e., $P(x \mid k) = P(x^{\mathrm{prop}}) = \mathcal{N}(0, 1)$) for different anchors. We found that for a wide range of realistic time costs, the optimal number of adjustments (see Figure 4.4(a)) is much smaller than the number of adjustments that would be required to eliminate the bias toward the anchor. Consequently, for a wide range of time costs, the estimate obtained after the optimal number of adjustments is still biased toward the anchor (see Figure 4.4(b)). This is a consequence of the geometric convergence of the error (see Figure 4.2), which leads to quickly diminishing returns for additional adjustments. This is a general property of this rational model of adjustment that can be derived mathematically (Lieder et al., 2012).

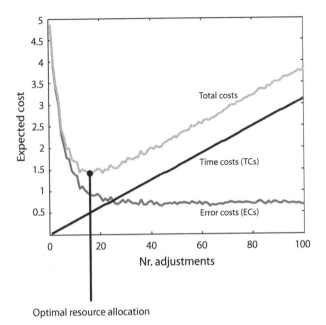

Optimal resource allocation

FIGURE 4.3. The expected value of the error cost $\text{cost}(x, \hat{x}_n)$ decays nearly geometrically with the number of adjustments n. While the decrease of the error cost diminishes with the number of adjustments, the time cost $\gamma \cdot t$ continues to increase at the same rate. Consequently, there is a point when further decreasing the expected error cost by additional adjustments no longer offsets their time cost. At that point, the total cost starts to increase. That point is the optimal number of adjustments $t^\star$.

4.1.4 Steps 4 and 5: Derive and test the model's behavioral predictions and determine if the model should be refined

Following the definition of the bias of an estimator in mathematical statistics, the anchoring bias can be quantified by $B_t(x, a) = \mathbb{E}[\hat{x}_t \mid x, a] - x$, where $\hat{x}_t$ is a participant's estimate of a quantity x after t adjustments, and a denotes the anchor. Figure 4.5 illustrates this definition and four basic ideas. First, the average estimate generated by anchoring-and-adjustment equals the anchor plus the adjustment. Second, the adjustment equals the relative adjustment times the total distance from the anchor to the posterior expectation. Third, adjustments tend to be insufficient, because the relative adjustment size is less than 1. Therefore, the average estimate usually lies between the anchor and the correct value. Fourth, because the relative adjustment is less than 1, the anchoring bias increases linearly with the distance from the anchor to the correct value.

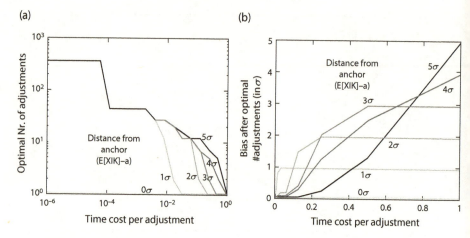

FIGURE 4.4. Optimal number of adjustments (a) and the bias after optimal number of adjustments (b) as a function of relative time cost and distance from the anchor ($1\sigma, 2\sigma, \ldots, 5\sigma$).

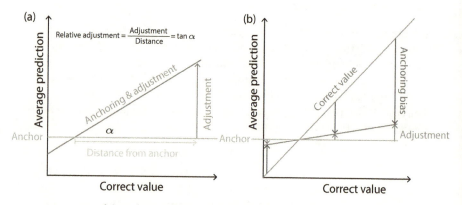

FIGURE 4.5. If the relative adjustment is less than 100%, then the adjustment is less than the distance from the anchor and the prediction is biased (Panel a), and the magnitude of the anchoring bias increases with the distance of the correct value from the anchor (Panel b).

The hypothesis that the mind performs probabilistic inference by sequential adjustment makes the interesting, empirically testable prediction that the less time and computation a person invests into generating an estimate, the more biased her estimate will be toward the anchor. As illustrated in the left panel of Figure 4.6, the relative adjustment (see Figure 4.5) increases with the number of adjustments. When the number of adjustments is zero, then the relative adjustment is zero, and the prediction is the anchor regardless of how far it is away from the correct value. However, as the number of

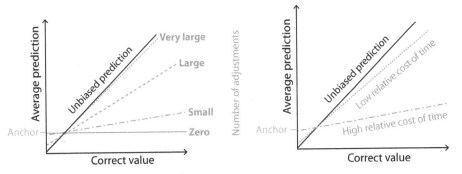

FIGURE 4.6. The number of adjustments increases the relative size of adjustments (left panel). As the relative cost of time increases, the number of adjustments decreases and so does the relative size of the adjustment (right panel).

adjustments increases, the relative adjustment increases, and the predictions become more informed by the correct value. As the number of adjustments tends to infinity, the average guess generated by anchoring-and-adjustment converges to the expected value of the posterior distribution. The predictions of the resource-rational anchoring-and-adjustment model were evaluated in three steps.

First, we used this model to simulate people's judgments in previously conducted anchoring experiments and found that it captured a wide range of anchoring biases and how they are affected by numerous factors, including time pressure, the extremity of the anchor, uncertainty, knowledge, and people's motivation to be accurate (Lieder et al., 2018b). Second, we designed two experiments specifically to test the model's prediction that the anchoring bias should increase with time pressure but decrease with error cost (Lieder et al., 2018c). The first experiment confirmed this prediction in a task where people generated their own anchors (see Figure 4.7), and the second experiment confirmed it in a task where people's anchors were provided by leading questions. Third, we conducted formal model comparisons and found that the resource-rational anchoring-and-adjustment model explained the data significantly better than alternative models of anchoring or numerical estimation (Lieder et al., 2017a).

In the fifth and final step, we concluded that our resource-rational model of numerical estimation captured the known empirical phenomena well enough. We therefore did not undertake any refinements.

In summary, the findings of our resource-rational analysis suggested that at least some anchoring biases might reflect the rational use of finite time and bounded cognitive resources rather than profound irrationality.

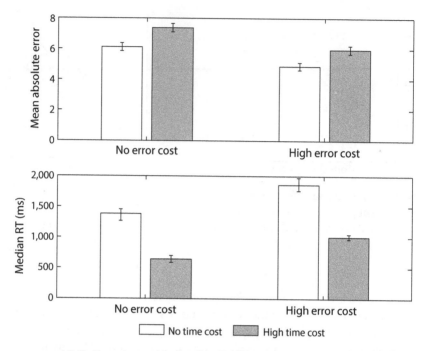

FIGURE 4.7. In Experiment 1 by Lieder, Griffiths, Huys, and Goodman (2018a), the mean absolute errors and reaction times as a function of time cost and error cost indicate an adaptive speed-accuracy trade-off.

4.2 Case study 2: A resource-rational perspective on availability biases in judgment and decision-making

In our second case study, we applied resource-rational analysis to elucidate how people make decisions under uncertainty (Lieder et al., 2018a), as when choosing between multiple risky gambles. To illustrate the methodology, we summarize this work in terms of the four steps of resource-rational analysis.

4.2.1 Step 1: Formalize the problem domain that a cognitive policy is optimized for

In the first step of our resource-rational analysis, we specified the goal of decision-making as choosing actions so as to maximize their expected utility. In other words, the function of decision-making is to select an action (a) whose expected utility is as high as possible. The expected utility of an action a integrates the probabilities $P(o \mid a)$ of its possible outcomes o with their utilities $u(o)$. Put more formally, the objective function of decision-making is

$$U(a) = \mathop{\mathbb{E}}_{o|a} [u(o)].$$ (4.4)

Note that we let the situation (s) be implicit for this case study.

Unlike simple laboratory tasks where each choice can yield only a few possible payoffs, many real-life decisions have infinitely many possible outcomes. As a consequence, the expected utility of action a becomes an integral:

$$V(a) = \mathop{\mathbb{E}}_{o|a}[u(o)] = \int p(o \mid a) \cdot u(o) \, do.$$ (4.5)

In general, it is intractable to compute the exact value of this integral. Thus, decision-makers have to choose actions in a way that approximately maximizes their expected utility without calculating its value exactly.

4.2.2 Step 2: Posit a class of cognitive policies that the brain could implement to solve this type of problem, along with the cost of the cognitive resources used by those systems

In the second step of our resource-rational analysis, we explored the implications of resource constraints on decision-making under uncertainty. Building on the evidence and arguments from Step 2 of our resource-rational analysis of numerical estimation (Section 4.1.2), we assumed that the mind's elementary computation is sampling. This general cognitive architecture is a promising starting point for modeling decision-making because sampling methods can provide an efficient approximation to integrals such as the expected utility in Equation 4.5 (Hammersley and Handscomb, 1964), and mental simulations of a decision's potential consequences can be thought of as samples.

We then expressed people's time and resource constraints as a limit on the number of samples, where each sample is a simulated outcome. Thus, the decision-maker's primary cognitive resource is a probabilistic simulator of the environment. The decision-maker can use this resource to anticipate some of the many potential futures that could result from taking one action versus another, but each simulation takes a non-negligible amount of time. Since time is valuable and the simulator can perform only one simulation at a time, the cost of using this cognitive resource is thus proportional to the number of simulations (i.e., samples).

Importance sampling is a popular sampling algorithm in computer science and statistics (Hammersley and Handscomb, 1964; Geweke, 1989) with connections to both neural networks (Shi and Griffiths, 2009) and psychological process models (Shi et al., 2010). It estimates a function's expected value with respect to a probability distribution p by sampling from an importance distribution q and correcting for the difference between p and q by down-weighting samples that are less likely under p than under q and up-weighting samples that are more likely under p than under q. Concretely, *self-normalized importance sampling* (Robert and Casella, 2009) draws s samples $x_1, \dots, x_s$ from a distribution q, weights the function's value $f(x_j)$ for each sample x_j by the **weight** $w_j = \frac{p(x_j)}{q(x_j)}$, and then normalizes its estimate by the sum of the weights:

$$x_1, \dots, x_s \sim q, \quad w_j = \frac{p(x_j)}{q(x_j)} \tag{4.6}$$

$$\underset{p}{\mathbb{E}}[f(x)] \approx \hat{E}_{q,s}^{\text{IS}} = \frac{1}{\sum_{j=1}^{s} w_j} \cdot \sum_{j=1}^{s} w_j \cdot f(x_j). \tag{4.7}$$

With finitely many samples, this estimate is generally biased. We hypothesized that the brain uses a strategy similar to importance sampling to approximate the expected utility gain $\mathbb{E}_{p(o|a)}[\Delta u(o)]$ of taking action a by

$$\overline{\Delta u}_{q,s}^{\text{IS}}(a) = \frac{1}{\sum_{j=1}^{s} w_j} \sum_{j=1}^{s} w_j \cdot \Delta u(o_j), \quad o_1, \dots, o_s \sim q, \tag{4.8}$$

and then approximating the optimal decision $a^\star = \arg\max_a \mathbb{E}_{p(o|a)}[\Delta u(o)]$ by $\hat{a}^\star = \arg\max_a \overline{\Delta u}_{q,s}^{\text{IS}}(a)$.

Note that importance sampling is a family of algorithms that differ in which importance distribution q they sample from. Therefore, Equation 4.8 doesn't define a single decision mechanism, but a class of decision mechanisms that differ in how likely different outcomes are to be considered. For simplicity, our resource-rational analysis assumes that the class of cognitive processes that the brain can implement to choose between two actions (i.e., Π) is equivalent to the class of decision procedures that chooses the action whose expected utility is higher according to some version of importance sampling.

While the first case study analyzed how much people should think, this case study focuses on what people should think about. Therefore, rather than finding the number of samples that achieves the best trade-off between the expected utility of the resulting decision and the cost of cognition, this analysis focuses on which potential outcomes people should simulate. For simplicity, we therefore don't include the number of samples as an additional dimension along which the potential decision mechanisms can differ.

We assume that for each of these mechanisms, the cognitive cost of reaching a decision comprises the cognitive cost of simulating a number of possible outcomes and then deriving a decision from the outcomes of those simulations. We assume that these costs are independent of how the outcomes are simulated.

Each importance distribution q yields a different estimator, and two estimators may recommend opposite decisions. Thus, in the third step of our resource-rational analysis, we investigated which distribution q yields the best decisions.

4.2.3 Step 3: Identify the cognitive policy that maximizes expected utility minus expected opportunity cost

If a decision has to be based on only a few simulated outcomes, then what is the optimal way to select them? Formally, the agent's goal is to maximize the expected utility gain of a decision made from only s samples. The agent always chooses the option whose expected utility it expects to be higher. Therefore, the agent will always choose the optimal action when the error in its estimate of the difference in expected utility is smaller than the true difference. Therefore, the absolute value of the error in the agent's estimate is an upper bound on the utility foregone by deciding based on the sample-based estimate rather than the true expected utilities. We can, therefore, approximate the resource-rational sampling-based decision mechanism by minimizing this upper bound. This can be approximated by minimizing the expected value of the squared error of $\overline{\Delta U}_{q,s}^{IS}$, which is the sum of its squared bias and its variance (Hastie et al., 2009). As the number of samples s increases, the estimate's squared bias decays much faster than its variance. Therefore, as the number of samples s increases, minimizing the estimator's variance becomes a good approximation to minimizing its expected squared error.

According to variational calculus, the importance distribution

$$q^{var}(o) \propto p(o) \cdot | \Delta u(o) - \mathbb{E}_p[\Delta U] | \qquad (4.9)$$

minimizes the variance of the utility estimate in Equation 4.8 (Geweke, 1998; Zabaras, 2010). This means that the optimal way to simulate outcomes in the service of estimating an action's expected utility gain is to overrepresent outcomes whose utility is much smaller or much larger than the action's expected utility gain. Each outcome's probability is weighted by how disappointing $(\mathbb{E}_p[\Delta u] - \Delta u(o))$ or elating $(\Delta u(o) - \mathbb{E}_p[\Delta u])$ it would be to a decision-maker anticipating to receive their decision's expected utility gain $(\mathbb{E}_p[\Delta u])$. Unlike in previous theories, this distortion was *not* introduced to describe human behavior but derived from the optimality principle of resource rationality: Importance sampling over-simulates extreme outcomes to minimize the mean-squared error of its estimate of the action's expected utility gain. It tolerates the resulting bias because it is more important to shrink the estimate's variance.

Unfortunately, importance sampling with q^{var} is intractable because it presupposes the expected utility gain $\mathbb{E}_p[\Delta u]$ that importance sampling is supposed to approximate. However, the average utility $\overline{\Delta u}$ of the outcomes of previous decisions made in a similar context could be used as a proxy for the expected utility gain $\mathbb{E}_p[\Delta u]$. That quantity has been shown to be automatically estimated by model-free reinforcement learning in the midbrain (Schultz et al., 1997). Therefore, people should be able to sample from the approximate importance distribution,

$$\tilde{q}(o) \propto p(o) \cdot \left| \Delta u(o) - \overline{\Delta u} \right|. \tag{4.10}$$

This distribution weights each outcome's probability by the extremity of its utility. Thus, on average, extreme events will be simulated more often than equally likely outcomes of moderate utility. We therefore refer to simulating potential outcomes by sampling from this distribution as **utility-weighted sampling** (UWS).

If the simulated outcomes were drawn representatively from the outcome distribution p, then one could obtain an unbiased estimate of the expected utility gain by simply averaging their utilities. However, since the simulated outcomes are drawn from the importance distribution $\tilde{q}$ rather than p, the agent has to correct for the difference between these two distributions by computing a weighted average instead (Equation 4.7). Concretely, each simulated outcome o_j should be weighted by the ratio $w_j = \frac{p(o_j)}{\tilde{q}(o_j)}$ of its probability under the outcome distribution p to its probability under the importance distribution $\tilde{q}$ from which it was sampled. Thus, the extreme outcomes that are overrepresented among the samples from $\tilde{q}$ will be down-weighted, whereas the moderate outcomes that are underrepresented among the samples from $\tilde{q}$ will be up-weighted. Because $\tilde{q}(o) \propto p(o) \cdot \left| \Delta u(o) - \overline{\Delta u} \right|$, the weight w_j of outcome o_j is $\frac{1}{\left| \Delta u(o) - \overline{\Delta u} \right|/z}$ for some constant z. Since the weighted average in Equation 4.7 is divided by the sum of all weights, the normalization constant z cancels out. Hence, given samples $o_1, \ldots, o_s$ from the utility-weighted sampling distribution $\tilde{q}$, the expected utility gain of an action or prospect can be estimated by

$$\overline{\Delta U}_{\tilde{q},s}^{\text{IS}} = \frac{1}{\sum_{j=1}^{s} 1/\left| \Delta u(o_j) - \overline{\Delta u} \right|} \cdot \sum_{j=1}^{s} \frac{\Delta u(o_j)}{\left| \Delta u(o_j) - \overline{\Delta u} \right|}. \tag{4.11}$$

If no information is available a priori, then there is no reason to assume that the expected utility gain of a prospect whose outcomes may be positive or negative should be positive, or that it should be negative. Therefore, in these situations, the most principled guess an agent can make for the expected utility gain $\mathbb{E}_p[\Delta u]$ in Equation 4.9—before computing it—is $\overline{\Delta u} = 0$. Thus, when the expected utility gain is not too far from zero, then the importance distribution q^{var} for estimating the expected utility gain of a single prospect can be efficiently approximated by

$$\tilde{q}(o) \propto p(o) \cdot |\Delta u(o)|. \tag{4.12}$$

This approximation simplifies the UWS estimator of a prospect's expected utility gain (Equation 4.11) into

$$\Delta \hat{U}^{\mathrm{IS}}_{\tilde{q},s} = \frac{1}{\sum_{j=1}^{s} 1/|\Delta u(o_j)|} \cdot \sum_{j=1}^{s} \mathrm{sign}\left(\Delta u(o_j)\right), \quad o_j \sim \tilde{q}, \tag{4.13}$$

where $\mathrm{sign}(x)$ is -1 for $x < 0$, 0 for $x = 0$, and $+1$ for $x > 0$.

Next, we translated the resource-rational mechanism for estimating expected utilities into a decision strategy. Many real-world decisions and most laboratory tasks involve choosing between two actions a_1 and a_2 with uncertain outcomes $O^{(1)} \in \{o_1^{(1)}, o_2^{(1)}, \ldots, o_{n_1}^{(1)}\}$ and $O^{(2)} \in \{o_1^{(2)}, o_2^{(2)}, \ldots, o_{n_2}^{(2)}\}$ that depend on the unknown state of the world. Consider, for example, the choice between two lottery tickets: the first ticket offers a 1% chance to win $1,000 at the expense of a 99% risk to lose $1 ($o^{(1)} \in \{-1, 1,000\}$) and the second ticket offers a 10% chance to win $1,000 at the expense of a 90% risk to lose $100 ($o^{(2)} \in \{-100, 1,000\}$). According to expected utility theory, one should choose the first lottery (taking action a_1) if $\mathbb{E}[u(o^{(1)})] > \mathbb{E}[u(o^{(2)})]$ and the second lottery (action a_2) if $\mathbb{E}\left[u(o^{(1)})\right] < \mathbb{E}\left[u(o^{(2)})\right]$. This is equivalent to taking the first action if the expected utility difference $\mathbb{E}\left[u(o^{(1)}) - u(o^{(2)})\right]$ is positive and the second action if it is negative. The latter approach can be approximated very efficiently by focusing computation on those outcomes for which the utilities of the two actions are very different and ignoring events for which they are (almost) the same. For instance, it would be of no use to simulate the event that both lotteries yield $1,000 because it would not change the decision-maker's estimate of the differential utility and thus have no impact on her decision. To make rational use of their finite resources, people should thus use utility-weighted sampling to estimate the expected value of the two actions' differential utility $\Delta u = u(o^{(1)}) - u(o^{(2)})$ as efficiently as possible. This is accomplished by sampling pairs of outcomes from the importance distribution

$$q_\Delta^\star(o^{(1)}, o^{(2)}) \propto p(o^{(1)}, o^{(2)}) \cdot \left| u(o^{(1)}) - u(o^{(2)}) - \mathbb{E}\left[\Delta u\right]\right|, \tag{4.14}$$

integrating their differential utilities according to

$$\Delta \hat{U}^{IS}_{q,s} = \frac{1}{\sum_{j=1}^{s} w_j} \sum_{j=1}^{s} w_j \cdot \left(u(o_j^{(1)}) - u(o_j^{(2)}) \right),$$

$$o_1, \ldots, o_s \sim q_\Delta \left(o^{(1)}, o^{(2)} \right), \tag{4.15}$$

and then choosing the first action if the estimated differential utility is positive, that is

$$\hat{a}^\star = \begin{cases} 1 \text{ if } \Delta \hat{U}^{IS}_{q,s} > 0 \\ 2 \text{ if } \Delta \hat{u}^{IS}_{q,s} < 0 \\ 1 \text{ with 50\% probability and 2 50\% probability if } \Delta \hat{U}^{IS}_{q,s} = 0. \end{cases} \tag{4.16}$$

Note that each simulation considers a pair of outcomes: one for the first alternative and one for the second alternative. This is especially plausible when the outcomes of both actions are determined by a common cause. For instance, the utilities of wearing a shirt versus a jacket on a hike are both primarily determined by the weather. Hence, reasoning about the weather naturally entails reasoning about the outcomes of both alternatives simultaneously and evaluating their differential utilities in each case (e.g., rain, sun, wind, etc.) instead of first estimating the utility of wearing a shirt and then starting all over again to estimate the utility of wearing a jacket.

Given that there is no a priori reason to expect the first option to be better or worse than the second option, $\mathbb{E}[\Delta u]$ is 0, and the equation simplifies to

$$q_\Delta(o^{(1)}, o^{(2)}) \propto p(o^{(1)}, o^{(2)}) \cdot \left| u(o^{(1)}) - u(o^{(2)}) \right|. \tag{4.17}$$

This distribution captures the fact that the decision-maker should never simulate the possibility that both lotteries yield the same amount of money—no matter how large it is. It does not overweight extreme utilities per se, but rather pairs of outcomes whose utilities are very different. Its rationale is to focus on the outcomes that are most informative about which action is best. For instance, in the example above, our UWS model of binary choice overweights the unlikely event in which the first ticket wins \$1,000 and the second ticket loses \$100. Plugging the optimal importance distribution (Equation 4.17) into the UWS estimate for the expected differential utility yields an intuitive heuristic for choosing between two options. Formally, the optimal importance sampling estimator for the expected value of the differential utility

$(\mathbb{E}\,[\Delta u])$ is

$$\Delta \hat{u}_{q,s}^{IS} = \frac{1}{\sum_{j=1}^{s} 1/\left| u\left(o_j^{(1)}\right) - u\left(o_j^{(2)}\right)\right|}$$

$$\cdot \sum_{j=1}^{s} \text{sign}\left(u\left(o_j^{(1)}\right) - u\left(o_j^{(2)}\right)\right), \quad o_j \sim q_\Delta, \qquad (4.18)$$

where sign(x) is $+1$ for positive x and -1 for negative x. If the heightened availability of extreme events roughly corresponded to the utility-weighted sampling distribution (Equation 4.17), then the decision rule in Equation 4.18 could be realized by the following simple and psychologically plausible heuristic for choosing between two actions:

1. Imagine a few possible events
   (e.g., 1. Ticket 1 wins and ticket 2 loses. 2. Ticket 2 wins and ticket 1 loses. 3. Ticket 1 winning and ticket 2 losing comes to mind again. 4. Both tickets lose.).
2. For each imagined scenario, evaluate which action would fare better
   (1. Ticket 1, 2. Ticket 2, 3. Ticket 1, 4. Ticket 1).
3. Count how often the first action fared better than the second one
   (three out of four times).
4. If the first action fared better more often than the second action, then choose the first action, else choose the second action (get ticket 1!).

4.2.4 Steps 4 and 5: Derive and test the model's behavioral predictions and determine if the model should be refined

In the fourth step of our resource-rational analysis, we evaluated our resource-rational model against empirical data. The UWS model predicts the well-known phenomenon that people overestimate the frequency of extreme events. We experimentally tested this prediction by asking people to judge the extremity and the relative frequency of mundane events, stressful life events, and lethal events. A significant rank correlation between participants' extremity judgments and the extent to which they overestimated each event's frequency confirmed this prediction (see Figure 4.8).

The UWS model can explain a wide range of previously reported biases in decisions from description, including the Allais paradox introduced in Chapter 2 (see Table 2.1). When the value of the irrelevant outcome (z) changes from \$2,400 to \$0, the preferences of UWS reverse in the same direction as people's preferences. For the first pair of lotteries $(z = 2,400)$, UWS

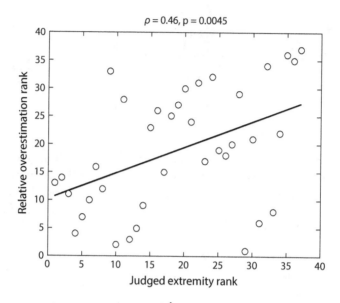

FIGURE 4.8. Relative overestimation $((\hat{f}_k - f_k)/f_k)$ increases with perceived extremity $(|u(o_k)|)$. Each circle represents one event's average ratings.

chooses the second lottery more often than the first. But for the second pair of lotteries $(z = 0)$, UWS chooses the first lottery more often than the second one. Why does this happen? UWS estimates the expected utility difference between choosing Lottery 1 and choosing Lottery 2 by importance sampling. In doing so, it overweights the event in which the utility difference between the two lotteries' outcomes $(o_1$ and $o_2)$ is most extreme $(\Delta u = u(o_1) - u(o_2))$. When $z = 2,400$, the second lottery yields \$2,400 for sure, whereas the first lottery has a 66% chance to yield \$2,400, a 33% chance to yield \$2,500, and 1% chance to yield nothing. So, in this case, the most overweighted event is the possibility that the first lottery yields $o_1 = 0$ and the second lottery yields $o_2 = 2,400$ $(\Delta U = -u(2,400))$. Consequently, the bias of UWS's estimate of the difference in expected utility is negative, and the first lottery appears inferior to the second. But when $z = 0$, then L_1 yielding $o_1 = 2,500$ and L_2 yielding $o_2 = 0$ $(\Delta U = +u(2,500))$ becomes the most overweighted event, and this makes the first lottery appear superior.

The UWS model also predicts Kahneman and Tversky's famous findings about how people's risk preferences differ between gains versus losses (framing effects) and between likely versus unlikely gains/losses (Lieder et al., 2018a). These findings are commonly explained by prospect theory (Kahneman and Tversky, 1979). According to prospect theory, the distortion of an outcome's probability is independent of its utility. UWS provides an alternative explanation, according to which this distortion is driven by the extremity of the outcome's utility.

Moreover, UWS can explain why people tend to remember the most extreme outcomes they encountered in a previous decision situation first (Madan et al., 2014) and why they overweight the most extreme previously experienced outcomes in decisions from experience (Ludvig et al., 2014).

In addition to confirming several predictions of the initial version of our UWS model, we also encountered a phenomenon that the model was unable to explain. Concretely, Ludvig et al. (2014) demonstrated that how much people overweight the more extreme potential outcome in one type of decision depends on whether it is more or less extreme than the potential outcomes in other previous decisions. We discovered that our initial version of UWS, which used the standard utility function from prospect theory (Kahneman and Tversky, 1979), was unable to explain this finding (see Lieder et al., 2018a). Therefore, in the fifth and final step of our resource-rational analysis, we refined the model's assumption about the utility function according to more biologically plausible principles of efficient coding (Summerfield and Tsetsos, 2015). This more realistic utility function takes into account that the brain's limited resources constrain the mind's ability to accurately represent utilities to a limited range of possible values. As a result, the resource-rational representation of value is sensitive to the range of values that can occur across different decisions. Taking this into account allowed the refined model to explain why people overweight the more extreme value less in environments where even more extreme values occur in other decisions.

In summary, our resource-rational analysis of decision-making suggested that at least some instances of the availability bias in judgment and decision-making likely reflect the rational use of limited information, finite time, and bounded cognitive resources.

4.3 Other resource-rational models for understanding heuristics and biases

Our case studies are just two examples of many applications of resource-rational analysis to heuristics and biases in human judgment and decision-making. In addition to the anchoring bias and the availability biases analyzed in the two case studies, human decision-making is also subject to many other biases and inconsistencies.

The resource-rational explanation of the Allais paradox we reviewed above was significant because the Allais paradox violates one of the three fundamental axioms of rational choice, namely the axiom of independence of irrelevant alternatives. Another, more direct, violation of this axiom is the finding that people's preference between two good alternatives depends on the attributes of a third alternative that is inferior to both of them. Such effects are known as contextual preference reversals. There are multiple resource-rational explanations of this cognitive bias as a consequence of people's rational use of their inevitably imprecise and noisy percepts and mental representations of the options' attributes (Matêjka and McKay, 2015; Howes et al., 2016). For example, Howes et al. (2016) showed that optimally combining noisy estimates of expected utility with noisy observations of the ordering of the options with respect to each attribute produces context-dependent preference reversals. The noisy observations assumed by their model are an inevitable consequence of people's limited cognitive resources. People simply cannot perceive and represent the world with unlimited fidelity and infinite precision because their cognitive resources are limited. Therefore, the best they can do is to use the resulting noisy representations as rationally as possible, even when the resulting strategy violates the axiom of independence of irrelevant alternatives.

Another reason why people's choices violate the independence axiom is that how much value people assign to an outcome depends on what they compare it to. This is known as reference-dependent preferences (O'Donoghue and Sprenger, 2018). Many reference-dependent preferences result from a phenomenon known as **loss aversion** (Kahneman and Tversky, 1979). Loss aversion is the phenomenon that, for most decision-makers, losing x is much worse than not gaining x, even when their resulting wealth is exactly the same. As a consequence, most people would refuse to play a lottery with a 50/50 chance of either winning $10 or losing $6. Whether an outcome, such as paying $200 for a TV, is perceived as a loss or a gain depends on what people compare it to, such as the advertised price, is larger or smaller than that outcome. This makes people's preferences highly sensitive to their point of comparison, which is called the reference point. Several researchers have proposed that these effects could result from the rational allocation of limited resources to the representation of gains versus losses (Woodford, 2012; Bhui and Gershman, 2018; Polanía et al., 2019; Bhui et al., 2021). The first premise of this idea is that constructing and maintaining precise representations with high fidelity requires costly cognitive resources. The second premise is that the brain's architecture for representing value is optimized to produce precise representations of values that are common at the expense of imprecise representations of values that are uncommon. Consistent with this premise, Polanía et al. (2019) found that people's preferences are more consistent for items whose values are common than for items whose values are rare.

Bhui and Gershman (2018) showed that a particular implementation of these principles can explain loss aversion. Building on earlier work by Stewart et al. (2006), they proposed that rational use of limited resources produces a representation that captures differences between losses (e.g., having to pay $4.95 versus having to pay $6.69 for a dessert) more precisely than differences between gains (e.g., winning $4.95 versus winning $6.69 in a lottery) because people make many more decisions in which they have to compare prices for purchasing decisions than decisions between options generating different amounts of income (e.g., deciding between job offers). According to this explanation, people are loss-averse because their mental representation of losing $6 is further away from their mental representation of losing/gaining nothing than their mental representation of gaining $10. This suggests that the numerous violations of the independence axiom that result from loss-aversion and reference-dependent preferences might be a consequence of the rational allocation of people's limited cognitive resources for constructing and maintaining precise representations of value.

Another example of people's decisions being influenced by supposedly irrelevant contextual factors is that consumers often fail to adjust how they spend their money when the balance on their bank account drops, or better options become available (Sims, 1998, 2003; Bhui et al., 2021). The resource-rational explanation for this apparently irrational behavior is that it could be exceedingly costly to invest all the attentional resources required to constantly monitor and precisely represent all aspects of the decision environment, including the current balance of one's bank account. So, people instead maintain imprecise representations of the decision environment that are not very responsive to small or moderate changes.

In addition to the axiom of independence, people's preferences also violate another axiom of rational choice, namely the axiom of transitivity. That is, under certain conditions, people prefer A to B and prefer B to C, but they don't prefer A over C. The resource-rational model by Tsetsos et al. (2016) suggests that these violations result from people mitigating the inevitable noise in their observations and representations of the alternatives' attributes and value in a way that improves the expected utility of their choices. Therefore, these violations of the axiom of transitivity might be consistent with the rational use of limited cognitive resources.

People's mental representations of outcome probabilities are also impacted by the noisiness of perception and the limited fidelity of mental representations. Khaw et al. (2021) showed that the optimal use of such imprecise representations gives rise to a bias in human decision-making known as **risk aversion**. Risk aversion is the phenomenon that people often decline risky gambles with positive expected utility in favor of less risky gambles with a lower expected utility.

Resource-rational analysis has also been applied to explain the **present bias**, which is people's irrational tendency to discount non-immediate costs and benefits (Gershman and Bhui, 2020; Gabaix and Laibson, 2017; Gabaix, 2023). According to these rational models, the present bias occurs partly because mental simulations of the distant future produce more uncertain, and hence less valuable, information per unit of mental effort than mental simulations of the immediate consequences (Gabaix and Laibson, 2017; Gershman and Bhui, 2020; Gabaix, 2023).

Besides cognitive biases, which are systematic errors, resource-rational models can also explain random variability in people's choices, such as why the same person might make different choices in two identical instances of the same decision (Bhui et al., 2021; Sims, 2003; Woodford, 2014; Matějka and McKay, 2015). The resource-rational explanation for this inconsistency is that representing the values of the options precisely enough to reliably reach the same decision every time is not always worth the resulting gain in reward.

Another argument against human rationality is the prevailing charge that people generally think too little (Kahneman, 2011). This charge is based on findings demonstrating that people often make mistakes that could have been avoided by further deliberation. However, this interpretation ignores the opportunity cost of people's time. Every minute a person spends on deliberating about what to do takes away from the time they have to do other things and deliberate about subsequent decisions. Research formally investigating how people allocate their limited time across multiple decisions generally found that the amount of time people invest into simple decisions tends to be near-optimal or too high (Dickhaut et al., 2009; Holmes and Cohen, 2014; Woodford, 2014; Oud et al., 2016). These findings suggest that the problem with human decision-making is not that we always invest too little time, but rather that we sometimes fail to notice when additional deliberation is *not* worth our time. This may be different for complex decisions.

Many resource-rational models explain the systematic and unsystematic errors of human decision-making in terms of rational heuristics (Lieder and Griffiths, 2020b). In many cases, the rational process models produced by a resource-rational analysis coincide with known heuristics. As a result, several heuristics that were once considered irrational, such as the availability heuristic and anchoring-and-adjustment, have been vindicated. In other cases, resource-rational analysis provided formal, mathematical support for earlier recommendations that people should use simple heuristics. For instance, a model proposed by Caplin et al. (2011) justified the **satisficing** heuristic proposed by Simon (1955). A resource-rational analysis by Callaway et al.

(2022b) validated the heuristic planning strategies proposed by Newell et al. (1972), and a resource-rational analysis by Krueger et al. (2024) validated the Take-The-Best heuristic proposed by Gigerenzer and Goldstein (1996).

These and other resource-rational analyses of the representations and strategies of judgment and decision-making jointly suggest that many inconsistencies and biases in human decision-making can be reinterpreted as manifestations of the rational use of limited cognitive resources. While this section could only provide a high-level overview of these additional resource-rational analyses, we will take a deeper dive into our resource-rational analysis of the heuristics people use to choose between alternatives (Krueger et al., 2024) and people's planning strategies (Callaway et al., 2022b) in Chapter 6.

4.4 Implications for the debate about human rationality

The standard picture of rationality posits that people should reason according to the rules of logic and probability theory and act so as to maximize their expected utility. This account was the foundation of Kahneman and Tversky's highly influential research program on heuristics and biases. Their discovery that human judgment and decision-making violate the traditional account of human rationality sparked an ongoing debate. The standard picture of human rationality was a fire waiting to happen; and the idea that people reason by applying the rules of logic and probability theory, and make calculated decisions that maximize expected utility, has since been reduced to ashes.

However, adherence to the rules of logic, probability theory, and expected utility theory is a flawed notion of rationality, because it ignores people's computational limitations. What, then, is to blame for people's violations of the normative principles? Is it the limitations of the human mind, or, rather, the limitations of our theories of rationality? The approach introduced in Chapter 3 addressed several key limitations of the standard picture of rationality by proposing a qualitatively different view of what it means to be rational. Granting that people have only limited cognitive resources and finite time to tackle the many millions of big and small decisions they have to make throughout life, resource-rational analysis focuses on modeling the optimal use of finite time and limited cognitive resources.

This new perspective challenges the pervasive interpretation that violating the rules of logic, probability theory, and expected utility theory is a sign of irrationality. It thereby insulates the question of human rationality from the empirical demonstrations that first set the standard picture of rationality on fire. Rather than being proof of human irrationality, the anchoring bias, availability biases in memory recall, judgment, and decision-making, and numerous other cognitive biases, could reflect the rational use of limited resources. This shows that resource-rational analysis has the potential to reconcile cognitive biases with rational models by building bridges from rational theories of behavior to heuristics and other psychological process models (Griffiths et al., 2015). Overall, research based on resource-rational analysis paints a brighter and more nuanced picture of human rationality than the traditional literature on heuristics and biases. From this new perspective, there is still hope for human rationality.

These results suggest that it may be time to revisit the debate about human rationality, with a different standard for human judgment and decision-making. Redefining human rationality in this way calls for redefining "cognitive biases" as the systematically suboptimal use of finite time and limited cognitive resources, which could manifest as thinking too much, thinking too little, or thinking ineffectively. Many violations of the standard picture of rationality will likely turn out not to be cognitive biases when analyzed from the resource-rational perspective. This means that the implications of virtually all previous findings on heuristics and biases will have to be re-evaluated. Very few studies have explicitly compared people's cognitive strategies against resource-rational heuristics. The question of human (ir)rationality thus remains unanswered.

One can revisit the rationality debate by going through the list of demonstrated violations of logic, probability theory, and expected utility and determine which of them are (in)compatible with resource rationality. The research surveyed in this chapter took that approach. Another approach is to derive and test qualitative predictions produced by resource-rational analysis and then evaluate them in human experiments. This could be seen as rebooting the research program of Kahneman and Tversky using a more accurate standard for human rationality. We believe there is great value in identifying cases where human cognition systematically deviates from resource-rational strategies. These genuine irrationalities would be promising targets for interventions aimed at improving human judgment and decision-making (see Chapter 8).

In this chapter, we showed how existing heuristics can be understood from the perspective of resource-rational analysis. In the next two chapters, we turn to two other fundamental problems that arise when we start thinking about human cognition in terms of a set of heuristics. First, how should we choose between those heuristics when faced with a new problem to solve or a decision to make? This is the focus of Chapter 5. Second, how can we discover new heuristics, either as researchers or as decision-makers? This is the focus of Chapter 6. Together, these results begin to show how people adapt their cognitive strategies to the structure of the problems they encounter.

5

Strategy Selection, Metareasoning, and Learning to Be Rational

To succeed in life, we have to solve a wide range of problems that place very different demands on us: sometimes we have to think fast, and sometimes we have to think slowly (cf. Kahneman, 2011). For instance, avoiding a car accident requires a split-second decision, whereas founding a successful start-up requires investing a lot of time into anticipating the future and weighing potential outcomes appropriately. No single decision mechanism works well across all situations. To meet the wide range of demands posed by different decision problems, the human brain, it has been proposed, is equipped with multiple decision systems (Dolan and Dayan, 2013) and decision strategies (Payne et al., 1988). Dual-process theories are a prominent example of this perspective (Evans and Stanovich, 2013; Evans, 2003; Kahneman, 2011). The coexistence of multiple alternative strategies is not specific to decision-making. People also appear to possess multiple strategies for inference, memory, self-control, problem-solving, and mental arithmetic, to name just a few.

The availability of multiple strategies that are applicable to the same problems raises the question of how people decide when to use which strategy. The fact that so many different strategies have been observed under different circumstances shows that people's strategy choices are highly variable and contingent on the situation and the task (Beach and Mitchell, 1978; Fum and Del Missier, 2001; Payne, 1982; Payne et al., 1988). Overall, the contingency of people's strategy choices appears to be adaptive. Even though under certain circumstances, people have been found to use heuristics that cause systematic errors (Ariely, 2009; Sutherland, 1992), their strategies are typically well adapted to the problems to which they are applied (Anderson, 1990; Braver, 2012; Bröder, 2003; Fum and Del Missier, 2001; Payne et al., 1993). For instance, Payne and colleagues found that when the probabilities of alternative outcomes fall off quickly, decision-makers employ frugal heuristics that prioritize the most probable outcomes at the expense of less probable ones. Similarly, decision-makers select fast heuristics when they are under time pressure but more accurate ones when they are not (Payne et al., 1988).

So, how can we explain the variability, task- and context-dependence, and adaptiveness of people's strategy choices? In other words, how does the human mind solve the problem of strategy selection? In this chapter, we formalize the strategy selection problem, derive a rational strategy selection mechanism, and show that it can explain a wide range of empirical phenomena, including the variability, contingency, and change of strategy selection across multiple domains. Our theory adds an important missing piece to the puzzle of bounded rationality by specifying when people should use which heuristic, and our findings reconcile the two poles of the debate about human rationality by suggesting that people gradually learn to make increasingly more rational use of their fallible heuristics.

5.1   The strategy selection problem

Each environment E can be characterized by the relative frequency P_E with which different kinds of problems occur in it. In most environments, these problems are so diverse that none of people's strategies can achieve the optimal speed-accuracy trade-off on all of them. Optimal performance in such environments requires selecting different strategies for different types of problems. For instance, it may be beneficial to select the satisficing heuristic in low-stakes decisions with many alternatives, such as choosing a dish at a restaurant, but to calculate each option's expected value in a high-stakes decision between two potential investments.

One way to achieve this would be to learn the optimal strategy for each problem separately through trial and error. This approach is unlikely to succeed in complex environments where no problem is exactly the same as any of the previous ones. Hence, in many real-world environments, learning about each problem separately would leave the agent completely unprepared for problems it has never seen before. This can be avoided by exploiting the fact that each problem has perceivable features $f_1, \ldots, f_K$ that can be used to predict the performance of candidate strategies from their performance on previous problems. For instance, for decisions with uncertain outcomes, these features may include the number of options, the spread of the outcome probabilities, and the range of possible payoffs.

The value of applying a cognitive strategy h (e.g., satisficing) to a particular problem is a special case of the value of cognition (VOC) introduced in Chapter 3. Like the VOC of performing a single cognitive operation, it depends on the expected reward and the expected cost of executing the strategy, that is,

$$\text{VOC}(\text{problem}, h) = \mathbb{E}[u(h(\text{problem}); \text{problem}) - \text{cost}(h, \text{problem})],$$

where $h(\text{problem})$ is the action the potentially stochastic strategy h selects for a problem, $u(a)$ denotes the utility of taking action a, and $\text{cost}(h, \text{problem})$ is the computational cost of executing strategy h on that problem. In the following, we will assume that the computational cost is driven primarily by the (cognitive) opportunity cost of the strategy's execution time $T(h, \text{problem}^{(i)})$, that is,

$$\text{cost}(h, \text{problem}^{(i)}) = \gamma \cdot T(h, \text{problem}^{(i)}).$$

The problem of optimal strategy selection can be defined as finding a mapping $\pi : \mathcal{F} \mapsto \mathcal{H}$ from feature vectors $(\mathbf{f}^{(i)} = (f_1(\text{problem}^{(i)}), \ldots, f_K(\text{problem}^{(i)})) \in \mathcal{F})$ to strategies $(h \in \mathcal{H})$ that maximizes the expected VOC of the selected strategy across all problems the environment might present. Formally, we can define the strategy selection problem as

$$\underset{m}{\arg\max} \sum_{\text{problem} \in \mathcal{P}} P_E(\text{problem}) \cdot \text{VOC}(\text{problem}, m(\mathbf{f}(\text{problem}))),$$

where $\mathcal{P}$ is the set of problems that can occur. This could, for instance, be the set of all decisions a person could potentially face in their lifetime.

Critically, the VOC of each strategy depends on the problem, but the strategy has to be chosen entirely based on the perceivable features $\mathbf{f}$, and the strategy selection mapping m has to be learned from experience. In machine learning, these kinds of problems are known as contextual multi-armed bandits (May et al., 2012). Two critical features of this class of problems are that they impose an exploration-exploitation trade-off and require generalization. In the next section, we will leverage these insights to derive a rational strategy selection learning mechanism.

The experience gained from applying a strategy h to a problem with perceivable features $\mathbf{f}$ and observing an outcome with utility u after executing the strategy for t units of time can be summarized by the tuple $(\mathbf{f}, h, u, t)$. This could, for instance, include the disappointment of finding out that all restaurants are already closed (u) after having wasted three hours (t) on comparing *all* restaurants on their average wait time (h) when you still had 327 options (f_1) at 9 pm (f_2). This experience, along with many others, would likely inform how you approach similar decisions in the future.

People's experience after the first n problems can be summarized by

$$e_n = \left((\mathbf{f}^{(1)}, h^{(1)}, u^{(1)}, t^{(1)}), \ldots, (\mathbf{f}^{(n)}, h^{(n)}, u^{(n)}, t^{(n)})\right),$$

where $\mathbf{f}^{(i)}, h^{(i)}, u^{(i)}, t^{(i)}$ are the feature vector of the ith problem, the strategy that was applied to it, and the resulting utility and execution time respectively. Strategy selection learning induces a sequence $m^{(1)}, m^{(2)}, \ldots, m^{(N)}$ of strategy selection mappings that depends on the agent's experience (e_n) and its strategy selection learning mechanism $l : \mathcal{E} \mapsto \mathcal{M}$, where $\mathcal{E}$ is the set of possible sequences of experiences and $\mathcal{M}$ is the set of possible strategy selection mechanisms. With this notation, we can express the agent's performance on the nth problem by

$$\text{VOC}(\text{problem}^{(n)}, m^{(n)}(\mathbf{f}^{(n)})),$$

where $\pi^{(n)}(\mathbf{f}^{(n)})$ is the strategy the agent selects for the nth problem, and the strategy selection mapping $\pi^{(n)}$ is $l(e^{(n-1)})$. Since the problem is sampled at random, the expected performance at time step n is

$$V_n(l) = \mathop{\mathbb{E}}_{\text{problem}^{(n)} \mid E} \left[\text{VOC}\left(\text{problem}^{(n)}, m^{(n)}(\mathbf{f}^{(n)})\right) \mid m^{(n)} = l(e_{n-1}) \right].$$

$$(5.1)$$

If the agent solves N problems before it runs out of time, its total performance is

$$V_{\text{total}}(l) = \mathbb{E}\left[\sum_{n=1}^{N} V_n(l) \right].$$

$$(5.2)$$

Using this notation, we can define the optimal strategy selection learning mechanism l^* as the one that maximizes the agent's total expected value of computation across all possible sequences of problems, that is,

$$l^* = \arg\max_l V_{\text{total}}(l).$$

This concludes our analysis of the problem that strategy selection (learning) has to solve. We will now use this analysis as a starting point for deriving a rational strategy selection learning mechanism.

5.2 The optimal solution to the strategy selection problem

The adaptiveness of people's strategy choices appears to increase with experience. For instance, as children gain more experience with mental arithmetic, they gradually learn to choose effective and efficient strategies more frequently

(Siegler, 1999). In adults, adaptive changes in strategy selection have been observed on much shorter time scales. For instance, adults have been found to adapt their decision strategy to the structure of their decision environment within minutes as they repeatedly choose between different investments based on multiple attributes (Rieskamp and Otto, 2006). In a decision environment where the better investment option is determined by a single attribute, people learn to use a fast-and-frugal heuristic that ignores all other attributes. However, when the decision environment does not have that structure, people learn to integrate multiple attributes. This suggests that people approximate rational strategy selection by learning what strategy is resource-rational in what situations.

Our computational-level analysis identified that a general strategy selection learning mechanism should be able to transfer knowledge gained from solving one problem to new problems that are similar. In the reinforcement learning literature, generalization is typically achieved by parametric function approximation (Sutton and Barto, 2018). The simplest version of this approach is to learn the coefficients of a linear function predicting the value of a state from its features. Such linear approximations require minimal effort to evaluate and can be learned very efficiently. We therefore propose that people learn an internal predictive model that approximates the value of applying a strategy h to a problem by a weighted average of the problem's features $f_1(\text{problem}), \ldots, f_n(\text{problem})$:

$$\text{VOC}(\text{problem}, h) \approx \sum_{k=1}^{n} w_{k,h} \cdot f_k(\text{problem}). \tag{5.3}$$

This approximation is easy to evaluate, but it is not clear how it can be learned, given that the VOC cannot be observed directly. However, when the strategy h generates a decision, the VOC can be decomposed into the utility of the decision's outcome and the cost of executing the strategy. Assuming that the cost of executing the strategy is proportional to its execution time, the VOC can be approximated by

$$\text{VOC}(\text{problem}, h) \approx \mathbb{E}[U \mid \text{problem}, h] - \gamma \cdot \mathbb{E}[T \mid \text{problem}, h], \tag{5.4}$$

where U is the utility of the outcome obtained by following strategy h, γ is the agent's opportunity cost per unit time, and $\mathbb{E}[T \mid \text{problem}, h]$ is the expected execution time of the strategy h when applied to the problem.

Approximating the VOC thus becomes a matter of estimating three quantities: the expected utility of relying on the strategy, the opportunity cost per unit time, and the expected time required to execute the strategy. The agent can learn its opportunity cost γ by estimating its reward rate (Boureau et al., 2015; Niv et al., 2007). The utility of applying the strategy and its execution time T can be observed. Therefore, when the reward is continuous, it is possible to learn an efficient approximation to the VOC by learning linear predictive models of the utility of its decisions and its execution time and combining them according to

$$\text{VOC}(\text{problem}, h) \approx \sum_{k=1}^{n} w_{k,h}^{(R)} \cdot f_k(\text{problem}) - \hat{\gamma} \cdot \sum_{k=1}^{n} w_{k,h}^{(T)} \cdot f_k(\text{problem}).$$

This equation is a special case of the general approach specified in Equation 5.3. When the outcome is binary, then the predictive model of the reward takes the form

$$P(o = 1 \mid h, \text{problem}) = \frac{1}{1 + \exp\left(-\sum_{k=1}^{n} w_{k,h}^{(R)} \cdot f_k(\text{problem})\right)}.$$

We model the agent's estimate of its opportunity cost γ by the posterior mean $\mathbb{E}[\bar{r} \mid t_{\text{total}}, r_{\text{total}}]$ of its reward rate $\bar{r}$ given the sum of rewards r_{total} that the agent has experienced and the time since the beginning of the experiment (t_{total}).

Our theory covers learning and strategy selection. To simulate learning, the agent's belief about the reward rate and the feature weights in the predictive model of a strategy's accuracy and execution time are updated by Bayesian learning every time it is executed. The belief about the reward rate $\bar{r}$ is updated to $P(\bar{r} \mid r_{\text{total}}, t_{\text{total}})$. The weights of the execution time model are updated by Bayesian linear regression. The weights of the reward model are updated by Bayesian logistic regression if the reward is binary (i.e., correct versus incorrect), or by Bayesian linear regression when the reward is continuous (e.g., monetary). Lastly, our model learns which features are relevant for predicting the most recent strategy's execution time and reward by performing Bayesian model selection.

The second component of our model is strategy selection. Given the learned predictive models of execution time and reward, the agent could predict the expected VOC of each available strategy and select the strategy with the highest expected VOC. While this approach works well when the agent has already learned a good approximation to the VOC of each strategy, it ignores the value of learning about strategies whose performance is still uncertain. Hence, always using the strategy that appears best could prevent the agent from discovering that other strategies work even better. Yet, on average, strategies that appear suboptimal will choose worse actions than the strategy that appears best. This problem recapitulates the well-known exploration-exploitation dilemma in reinforcement learning. To solve this problem, our model selects strategies by Thompson sampling (May et al., 2012): for each strategy h, our model samples estimates $\tilde{w}_{k,h} = (\tilde{w}_{k,h}^{(T)}, \tilde{w}_{k,h}^{(R)})$ of the weights

$w_{k,h} = (w_{k,h}^{(T)}, w_{k,h}^{(R)})$ of the predictive models of execution time and reward from their respective posterior distributions, that is

$$\tilde{w}_{k,h}^{(T)} \sim p\big(w_{k,h}^{(T)} \mid e_{t-1,h}\big),$$

$$\tilde{w}_{k,h}^{(R)} \sim p\big(w_{k,h}^{(R)} \mid e_{t-1,h}\big),$$

where $e_{t-1,h}$ is the agent's past experience with strategy h at the beginning of trial t. From these weights $\tilde{w}$, our model predicts the VOC values of all strategies h by

$$\hat{V}_t(h, \text{problem}) = \sum_{k=1}^{n} \tilde{w}_{k,h}^{(R)} \cdot f_k(\text{problem})$$

$$- \mathbb{E}[\hat{\gamma} \mid h_t] \cdot \sum_{k=1}^{n} \tilde{w}_{k,h}^{(T)} \cdot f_k(\text{problem}),$$

where $\mathbb{E}[\hat{\gamma} \mid e_{t-1}]$ is the posterior expectation of the agent's reward rate given its past experience. Finally, our model selects the strategy h_t^* with the highest predicted VOC,

$$h_t^* = \arg\max_{h} \hat{V}_t(h, \text{problem}).$$

This concludes the description of our model.

Our proposal is similar to model-based reinforcement learning (Dolan and Dayan, 2013; Gläscher et al., 2010) in that it learns a predictive model. However, both the predictors and the predicted variables are different. While model-based reinforcement learning aims to predict the next state and reward from the agent's action (e.g., "Go left!"), our model learns to predict the costs and benefits of the agent's deliberation from the agent's cognitive strategy (e.g., planning four steps ahead versus planning only one step ahead). While model-based reinforcement learning is about the control of behavior, our model is about the control of mental activities that may have no direct effect on behavior. In brief, the main difference is that we have modeled metacognitive learning instead of stimulus-response learning. Despite this difference in semantics, the proposed learning mechanism is structurally similar to the semi-gradient SARSA algorithm from the reinforcement learning literature (Sutton and Barto, 2018).

As illustrated in Figure 5.1, our model's prediction mechanism could be approximated by a simple feedforward neural network. The first layer represents the input to the strategy selection network. The subsequent hidden layers extract features that are predictive of the strategy's execution time and accuracy. The second-last layer computes a linear combination of those features to predict the execution time and external reward of applying the

$$\mathrm{VOC}(h;f) \approx \mathbf{E}[R \mid h,f] - \gamma \cdot \mathbf{E}[T \mid h,f]$$

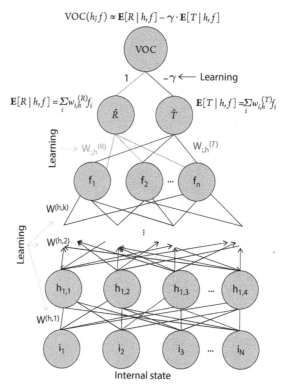

FIGURE 5.1. Our rational process model of strategy selection learning could be implemented in a simple feedforward neural network

strategy, and the final layer combines these predictions into an estimate of the VOC of applying the strategy in the current state. The weights of this network could be learned by a basic error-driven learning mechanism, and the features might emerge from applying the same error-driven learning mechanism to connections between the hidden layers (Mnih et al., 2015). With one such network per strategy, a simple winner-take-all network (Maass, 2000) could read out the strategy with the highest VOC. This neural network formulation suggests that a single forward pass through a few layers may be sufficient to compute each strategy's VOC. The action potentials and synaptic transmission required to propagate neural activity from one layer only take milliseconds. The winner-take-all mechanism for reading out the strategy with the highest VOC can be performed in less than 100 milliseconds (Oster et al., 2009). Hence, the brain might be able to execute the proposed strategy selection mechanism within fractions of a second.

Our simple model of strategy selection is based on an efficient approximation to the optimal solution prescribed by rational metareasoning. In contrast to previous accounts of strategy selection, our model postulates a more sophisticated, feature-based representation of the problem to be solved and a learning mechanism that achieves generalization. Instead of just learning about the reward that each strategy obtains *on average*, our model learns to predict each strategy's execution time and expected reward on each individual problem. Hence, while previous models learned which strategy works best on average, our model learns to predict which strategy is best for each individual problem. Whereas previous theories of strategy selection (Erev and Barron, 2005; Rieskamp and Otto, 2006; Siegler and Shipley, 1995; Siegler and Jeff, 1984; Siegler, 1988) rejected the ideal of a cost-benefit analysis as intractable, we propose that people learn to approximate it efficiently. Note, however, that the consideration of the cost of thinking (Shugan, 1980) is not the distinguishing feature of our model because costs can be incorporated into the reward functions of previous theories of strategy selection. Rather, the main innovation of our theory is that strategies are chosen based on the features of the problem to be solved. The remainder of this chapter shows that this allows our model to capture aspects of human cognition that were left unexplained by previous theories.

5.3 Testing the predictions of our rational metareasoning model

Previous models of strategy selection learning postulated that people learn to use the strategy that works best on average in a given environment (Shrager and Siegler, 1998; Erev and Barron, 2005; Rieskamp and Otto, 2006). By contrast, our rational metareasoning model learns to use the features of individual problems as relevant context. It can, therefore, learn to adaptively switch between different strategies within the same environment.

To test whether our model can provide a better account of human strategy selection than models that don't use features of individual decisions (context-free models), we designed a series of experiments in which learning to select strategies based on the features of individual problems leads to qualitatively different choices than learning to select the strategy that works best on average (Lieder and Griffiths, 2017).

5.3.1 Learning how to sort

In our first experiment, we taught 100 participants two unfamiliar strategies for sorting lists of numbers: cocktail sort and merge sort. Participants tried out these two sorting strategies on a series of lists that differed in their length and the degree to which they were already partially sorted (presortedness).

Neither of the two strategies was always faster than the other. Instead, the relative speed of each strategy depended on the list's length and presortedness. Cocktail sort is faster for lists that are largely presorted, whereas merge sort is faster for lists that aren't. After this initial learning phase, participants decided which of the two strategies they would apply to a series of 17 lists of different lengths and degrees of presortedness.

Overall, 83% of our participants chose merge sort more often when it was the superior strategy than when cocktail sort was the superior strategy, and vice versa. The high frequency of this adaptive strategy choice pattern provides strong evidence for the hypothesis that people's strategy choices are informed by features of the problem to be solved because it would be extremely unlikely otherwise. This finding was predicted by our rational metareasoning model of strategy selection, which achieved adaptive strategy selection in 70.5% of the simulations. These results support the conclusion that our participants' strategy choices were more adaptive than predicted by the three context-free models of strategy selection we tested our model against (Shrager and Siegler, 1998; Erev and Barron, 2005; Rieskamp and Otto, 2006). The frequency of adaptive strategy choices predicted by our rational metareasoning model was also significantly higher than for any alternative model. Like people, rational metareasoning selected merge sort for significantly more than half of the lists that were long and inverted but for significantly less than half of the lists that were long and presorted or short and presorted. None of the alternative models captured this pattern. For more details, see Lieder and Griffiths (2017) or Lieder et al. (2014b).

Our findings suggest that people leverage the features of individual problems to learn how to select strategies adaptively. The success of the rational metareasoning model and its evaluation against lesioned metareasoning models suggested that our hypothesis that people learn to predict the VOC of alternative strategies from the features of individual problems may be able to account for the adaptive flexibility of human strategy selection.

Unlike sorting strategies, most cognitive strategies operate on internal representations. In principle, strategies that operate on internal representations could be selected by a different mechanism than strategies that operate on external representations (e.g., numbers on a piece of paper). However, there are two reasons to expect our conclusions to transfer. First, people routinely apply strategies that they have applied to external objects to their internal representations of those objects. For instance, mental arithmetic is based on calculating with fingers. Thus, the strategies people use to order things mentally might also be based on the strategies they use to sort physical objects. Second, strategy selection can be seen as an instance of metacognitive control, and metacognitive processes tend to be domain-general. In the following sections, we show that our conclusions do indeed transfer to cognitive strategies that operate on internal representations.

5.3.2 Learning how to decide

People are known to use a wide repertoire of different heuristics to make decisions under risk (Payne et al., 1993). These strategies include fast-and-frugal heuristics, which, as the name suggests, perform very few computations and use only a small subset of the available information (Gigerenzer and Gaissmaier, 2011). For instance, the lexicographic heuristic (LEX) focuses exclusively on the most probable outcome that distinguishes between the available options and ignores all other possible outcomes. Another fast-and-frugal heuristic that people might sometimes use is Elimination-By-Aspects (EBA; Tversky, 1972). Here, we used the deterministic version of EBA described by Payne et al. (1988). This heuristic starts by eliminating options whose payoff for the most probable outcome falls below a certain threshold. If more than one option remains, EBA repeats the elimination process with the second most probable outcome. This process repeats until only one option remains, or all outcomes have been processed. After the elimination step, EBA chooses one of the remaining options at random. In addition to fast-and-frugal heuristics, people's repertoire also includes more time-consuming but potentially more accurate strategies, such as the weighted additive strategy (WADD). WADD first computes each option's expected value and then chooses the option whose expected value is highest.

In addition to gradually adapting their strategy choices to the structure of the environment (Rieskamp and Otto, 2006), people can also flexibly switch their strategy as soon as a different problem is presented. Payne et al. (1988) provided a compelling demonstration of this phenomenon in risky choice: participants chose between multiple gambles described by their possible payoffs and respective probabilities. There was a fixed set of possible outcomes that occurred with known probabilities, and the gambles differed in the payoffs they assigned to these outcomes. Participants were presented with four types of decision problems that were defined by the presence or absence of a time constraint (15 seconds versus none) and the dispersion of the outcomes' probabilities (low versus high); high dispersion means that some outcomes are much more probable than others, whereas low dispersion means that all outcomes are almost equally likely. Ten instances of each of the four problem types were intermixed in random order. The outcomes' payoffs ranged from $0 to $9.99, and their values and probabilities were stated numerically. Payne et al. (1988) used process tracing to infer which strategies their participants were using: the payoffs and their probabilities were revealed only when the participant clicked on the corresponding cell of the payoff matrix displayed on the screen, and all mouse clicks were recorded. This allowed Payne and colleagues to measure how often people used the fast-and-frugal heuristics (LEX and EBA) for different types of problems by the percentage of time they spent on the options' payoffs for the most probable outcome. For the expected-value strategy WADD, this proportion is only 25%, but for the fast-and-frugal heuristics LEX and EBA it can be up to 100%. The experiment revealed that people adaptively switch decision strategies in the absence of feedback: when the dispersion of outcome probabilities was high, people focused more on the most probable outcome than when all outcomes were almost equally probable. Time pressure also increased people's propensity for such selective and attribute-based processing. Thus, participants appeared to use fast-and-frugal heuristics more frequently when they had to be fast and when all but one or two outcomes were extremely improbable. This makes sense because the fast-and-frugal heuristics LEX and EBA are fast precisely because they focus on the most predictive attributes instead of integrating all attributes.[1]

We investigated whether rational metareasoning can account for people's adaptive flexibility in this experiment (Lieder and Griffiths, 2017). To do so, we simulated the experiment by applying our model to the selection between the ten decision strategies considered by Payne et al. (1988), including WADD and fast-and-frugal heuristics such as LEX and EBA. To simulate each strategy's execution time, we counted how many elementary operations (Johnson and Payne, 1985) it would perform on a given problem and assumed that each of them takes one second. This allowed us to simulate the effect of the time limit on a strategy's performance by having each strategy return its current best guess when it exceeds the time limit (Payne et al., 1988). For the purpose of strategy selection learning, our model represented each decision problem by five simple and easily computed features: the number of possible outcomes, the number of options, the number of inputs per available computation, the highest outcome probability, and the difference between the highest and the lowest payoff. Our model used these features to learn a predictive model of each strategy's relative reward

$$r_{\text{rel}}(s; o) = \frac{V(s(D), o)}{\max\limits_{a} V(a, o)},$$

---

1. The time and effort required to click on a cell also contributed to the cost participants sought to minimize in our experiment. However, it is well known that people don't integrate all information even when all costs are purely cognitive.

where $s(D)$ is the gamble that strategy s chooses in decision problem D, $V(c, o)$ is the payoff of choice c if the outcome is o, and the denominator is the highest payoff the agent could have achieved given that the outcome was o. To choose a strategy, the predicted relative reward $\hat{r}_{rel}$ is translated into the predicted absolute reward $\hat{r}$ by the transformation

$$\hat{r} = \min\{r_{min} + (r_{max} - r_{min}) \cdot \hat{r}_{rel}, r_{max}\},$$

where r_{min} and r_{max} are the smallest and the largest possible payoffs of the current gamble, respectively. The model then integrates the predicted absolute reward and the predicted time cost into a prediction of the strategy's VOC according to Equation 5.4 and chooses the strategy with the highest VOC as usual. The priors on all feature weights of the score and execution time models were standard normal distributions. The simulation assumed people knew their opportunity cost and did not have to learn it from experience. Rather than requiring the model to learn the time cost as outlined above, the opportunity cost was set to $7 per hour and normalized by the maximum payoff ($10) to make it commensurable with the normalized rewards.

To compare people's strategy choices to rational metareasoning, we performed 1,000 simulations of people's strategy choices in this experiment. In each simulation, we modeled people's prior knowledge about risky choice strategies by letting our model learn from ten randomly generated instances of each of the 144 types of decision problems considered by Payne et al. (1988). We then applied rational metareasoning with the learned model of the strategies' execution time and expected reward to a simulation of Experiment 1 from Payne et al. (1988). On each simulated trial, we randomly picked one of the four instances. We generated the payoffs and outcome probabilities according to the problem type: outcome distributions with low dispersion were generated by sampling outcome probabilities from standard uniform distribution and dividing them by their sums. Outcome distributions with high dispersion were generated by sampling the outcome probabilities sequentially such that the second-largest probability was at most 25% of the largest one, the third-largest probability was at most 25% of the second-largest one, and so on. Since the participants in this experiment received no feedback, our simulation assumed no learning during the experiment.

We found that rational metareasoning correctly predicted that time pressure and probability dispersion increase people's propensity to use the fast-and-frugal heuristics LEX and EBA. Time pressure increased the predicted frequency of fast, attribute-based processing by 30%, and high dispersion of the outcome probabilities increased the predicted frequency of fast, attribute-based processing by 44%. These results suggest that rational metareasoning can capture the adaptive flexibility of people's strategy choices not only for behavioral strategies that manipulate external representations but also for cognitive strategies that operate on internal representations. By contrast, none of the context-free models of strategy selection (SSL, RELACS, or SCADS) were able to explain these effects.

5.4 Learning to become more rational

In the previous sections, we presented a model according to which people's ability to adaptively select between alternative cognitive strategies is acquired through metacognitive learning. As the model learns to select strategies more adaptively, its decision-making becomes increasingly more resource-rational. Therefore, according to this model, strategy selection learning makes people increasingly more resource-rational. This section summarizes a series of experiments testing this prediction.

To test this hypothesis we used an experimental paradigm known as the Mouselab paradigm (see Figure 5.2). In our version of the Mouselab

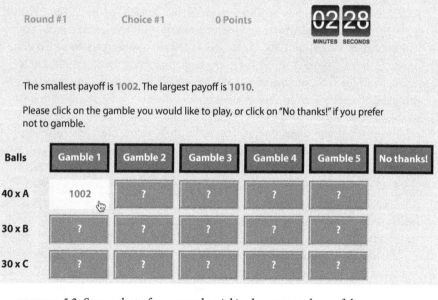

FIGURE 5.2. Screenshot of an example trial in the pretest phase of the experiment testing Prediction 1.

paradigm, participants choose between multiple gambles. Each gamble is defined by how many points it yields, depending on which type of ball is drawn from an urn. Participants know how many balls of each type are in the urn. Initially, they don't know how many points any gamble yields for any type of ball. To find out, they can click on the corresponding cell of a payoff matrix. The payoff matrix has one row for each type of ball and one column for each gamble. Once the participant clicks on a cell, the corresponding payoff is revealed. The participant can decide how many payoffs, if any, they want to inspect before selecting a gamble. Clicking on a cell and processing the revealed payoff comes at a cost: it takes time and cognitive effort. As a result, participants are strategic about which cells they click on. One can assume that each click corresponds to one or more cognitive operations that process the revealed payoff and incorporate it into the participant's mental representation of the value of the corresponding gamble. This allows the experimenter to infer how many cognitive operations the participant performed, which operations they performed, and in what order they performed them.

5.4.1 Learning from experience makes people increasingly more resource-rational.

According to our rational model of strategy selection learning, people acquire their capacity for adaptive strategy selection by learning an internal predictive model of each strategy's performance. This model predicts that people should gradually learn to perform more valuable computations and fewer computations whose costs outweigh their benefits. In other words, people should learn to make increasingly more rational use of their finite time and their bounded computational resources. This hypothesis makes four predictions:

1. People learn to perform fewer computations whose time cost outweighs the resulting gain in decision quality.
2. People learn to perform more computations whose expected gain in decision quality outweighs their time cost.
3. Ecological rationality increases with learning: people gradually learn to adapt their strategy choices to the structure of their environment.
4. Adaptive flexibility increases with learning: people learn to use different strategies for different kinds of problems.

We tested these predictions in a series of four experiments (Lieder and Griffiths, 2017).

PREDICTION 1: WHEN PEOPLE THINK TOO MUCH, THEY LEARN TO DELIBERATE LESS

The goal of the first experiment was to test our model's prediction that people will learn to deliberate less and decide more quickly when they are placed in an environment where the cost of deliberation outweighs its benefits.

The experiment was structured into three blocks: a pretest block, a training block, and a posttest block. Participants received feedback about the outcomes of their choices in the training block, but not in the pretest or the posttest block. Each block lasted four minutes. Participants received a financial bonus proportional to the number of points they earned in the task.

Figure 5.2 shows a screenshot of an example trial in the pretest phase. In each trial, participants were shown a number of gambles. They could either choose one of the gambles or skip the decision and move on to the next trial without receiving a payoff. As soon as the participant responded, the next trial was shown. The number of trials was solely determined by how quickly the participant responded on each trial. For some decisions, the optimal strategy was to pick one gamble at random without inspecting any payoffs. For other decisions, the best strategy was to skip the decision entirely. Crucially, the environment didn't include any decisions where the optimal strategy involved inspecting any of the payoffs.

Participants usually start off inspecting at least a couple of payoffs per decision. Therefore, our model predicted that participants would learn to inspect increasingly fewer payoffs and consequently earn increasingly more points per second. As predicted, as they progressed from the pretest block to the posttest block, participants learned to process less information and learned to skip significantly more decisions. Consequently, their average reward rate was significantly higher in the posttest block than in the pretest block.

This finding supports the view that people sometimes being cognitive misers is not necessarily evidence against human rationality. Instead, it could be an adaptation to an environment in which people's opportunity costs are high relative to the benefits of additional deliberation. According to our model, people become cognitive misers when the difference between the rewards for good versus bad performance is small compared to how much time it would take to perform better.

PREDICTION 2: WHEN PEOPLE THINK TOO LITTLE, THEY LEARN TO DELIBERATE MORE

The goal of the second experiment was to test our model's prediction that people learn to deliberate more when they initially think too little.

The experiment was divided into a series of five blocks. Each block lasted for a fixed amount of time. How many decisions a participant made in each block was entirely up to them. The faster they decided, the more decisions they could make. Each decision involved choosing between five gambles with four possible payoffs each. These payoffs were initially concealed, but participants could reveal any payoff they were interested in by clicking on it. We measured how much participants deliberated about each of these decisions by measuring how many of the 20 possible payoffs they inspected.

To create a situation where people think too little, we first put them in an environment where the optimal strategy is to make low-stakes decisions as quickly as possible. In that environment, participants had a fixed amount of time to make as many decisions as they wanted, and many of those decisions produced high payoffs regardless of what they chose. After letting participants make decisions in this environment for two sessions (Blocks 1 and 2), we changed the environment so that low-stakes decisions became the only opportunity to earn money (Blocks 3–5). According to our rational model, in Blocks 3–5, participants should adapt to this change in the environment by learning to deliberate increasingly more about their low-stakes decisions.

Confirming our model's prediction, the number of payoffs participants inspected in the low-stakes problems significantly increased from 5.97 in Block 3 to 6.42 in Block 5. This learning-induced change in people's decision strategies was accompanied by a significant improvement in participants' performance; they lost $2.00 in Block 3 but only $1.16 in Block 5.

PREDICTION 3: ECOLOGICAL RATIONALITY INCREASES WITH LEARNING

The third prediction of our model is that people adapt their strategy choices to the structure of their environment. To evaluate this prediction, we examined how people making a series of investment decisions adapt how much they rely on two different strategies: the comprehensive weighted additive strategy (WADD) versus the fast-and-frugal heuristic Take-The-Best (TTB). The WADD strategy sums up the weighted differences between the first option's rating and the second option's rating across all attributes (Tversky, 1969). TTB is the equivalent of the lexicographic heuristic for multi-attribute decisions: it chooses the option that is best on the most predictive attribute that differs between the options and that ignores all other attributes.

TTB works best when the attributes' predictive validities fall off so quickly that the recommendation of the most predictive attribute cannot be overturned by rationally incorporating additional attributes; environments with this property are called **non-compensatory**. By contrast, TTB can fail miserably when no single attribute reliably identifies the best choice by itself; and environments with this property are called **compensatory**. Thus, to adapt rationally to the structure of their environment, that is, to be **ecologically rational**, people should select TTB in non-compensatory environments and avoid it in compensatory environments.

Bröder (2003) found that people use TTB more frequently when their decision environment is non-compensatory. Rieskamp and Otto (2006) found that this adaptation might result from reinforcement learning. In their experiment, participants made 168 multi-attribute decisions with feedback. In the first condition, all decision problems were compensatory, whereas in the second condition, all decision problems were non-compensatory. To measure people's strategy use over time, Rieskamp and Otto (2006) analyzed their participants' choices on trials where TTB and WADD made opposite decisions. Participants in the non-compensatory environment learned to choose in accordance with TTB increasingly *more often*, whereas participants in the compensatory environment learned to do so increasingly *less* often.

To test whether our theory of strategy selection learning can explain these findings, we simulated the experiment using our rational metareasoning model (Lieder and Griffiths, 2015, 2017). Our simulation showed that rational metareasoning could explain people's ability to adapt their strategy choices to the structure of their environment, at least as well as the SSL model that Rieskamp and Otto (2006) developed to explain their findings (see Figure 5.3). When the decision environment was non-compensatory, then our model learned to use TTB and avoid WADD. However, when the decision environment was compensatory, our model learned to use WADD and avoid TTB. In addition, rational metareasoning model captured that people adapt their strategy choices gradually.

PREDICTION 4: ADAPTIVE FLEXIBILITY INCREASES WITH LEARNING

The fourth prediction of our model is that people learn to flexibly switch their strategies on a trial-by-trial basis to exploit the structure of individual problems.

Our simulations revealed that feature-based and context-free strategy selection learning predict qualitatively different effects of the relative frequency of compensatory versus non-compensatory decision problems. Concretely, the performance of context-free models of strategy selection learning drops rapidly as the decision environment becomes more heterogeneous: as the ratio of compensatory to non-compensatory problems approaches 50/50, the performance of the three context-free models (SSL, RELACS, and SCADS) drops to the level of chance. By contrast, the performance of feature-based

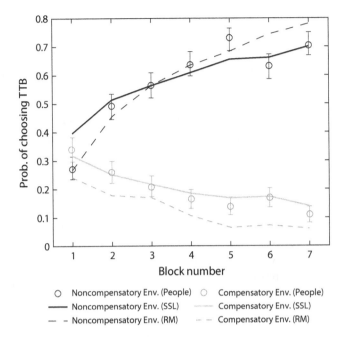

FIGURE 5.3. Fit of rational metareasoning model (RM) and a context-free model (SSL) to the empirical data by (Rieskamp and Otto, 2006).

strategy selection learning (rational metareasoning) is much less susceptible to this heterogeneity and stays above 60%. The reason is that rational metareasoning learns to use TTB for non-compensatory problems and WADD for compensatory problems, whereas context-free models learn to always use the same strategy. We, therefore, determined whether people rely on context-free or feature-based strategy selection by putting participants in a heterogeneous environment (Lieder and Griffiths, 2015).

In our experiment, 100 participants made 100 binary choices each. In each trial, participants chose between a new pair of investment options with uncertain outcomes. They could base their choice on one or more of several binary attributes and estimates of the attributes' relevance to the decision. After each choice, either option could turn out to be better.

Each participant earned a bonus of 1.25 cents for every correct decision and lost 1.25 cents for every incorrect decision. Critically, the ratio of compensatory to noncompensatory problems was 50/50 and the problems were chosen such that TTB and WADD make opposite decisions on every trial. In half of the trials, the option chosen by TTB turned out to be better, and in the other half, the option chosen by WADD turned out to be better. Therefore, always using TTB, always using WADD, choosing one of these two strategies at random, or using context-free strategy selection would perform at the chance level.

To determine the quality of people's strategy choices, we compared their decisions on each trial to those of the strategy that is more resource-rational for the problem presented on that trial. For compensatory trials, we evaluated people's choices against those of WADD. For non-compensatory trials, we evaluated them against TTB because TTB achieves a better time-accuracy trade-off in those decisions. On average, people's decisions agreed with those of the appropriate strategy in 76.2% of the trials. Moreover, their performance increased significantly from 70.4% in the first ten trials to 80.4% in the last ten trials. This level of performance could not have been achieved by context-free strategy selection, which performed at chance, but it is qualitatively consistent with feature-based strategy selection, which selected the correct strategy in more than 60% of the decision situations. All four predictions of our model were thus confirmed. Each time, we found that participants learned to become increasingly more resource-rational.

5.4.2 Additional evidence from strategy discovery and the plasticity of cognitive control

With extensive practice, cognition becomes increasingly more automatic and habitual. Although this automaticity often allows us to select resource-rational strategies with minimal cognitive effort, it isn't very flexible. Consequently, we sometimes encounter situations when we have to inhibit our automatic cognitive processes to override them with a deliberately chosen alternative. This cognitive ability is known as cognitive control. The classic example of a paradigm used to study cognitive control is the Stroop task. The Stroop task is to name the color of color words. To do this correctly, people have to inhibit the automatic cognitive process of word reading and override it with the deliberately chosen process of color naming. But how does the brain determine when to override the automatic process, and what deliberate strategy to override it with? This problem is, in many ways, analogous to the strategy selection problem. We, therefore, adopted our rational metareasoning model of strategy selection to the allocation of cognitive control (Lieder et al., 2018d). According to this model, people learn to override automatic processes when the resulting increase in accuracy is more valuable than the cost of exerting cognitive control. To achieve this, people learn to predict the net benefit of exerting cognitive control from observable features of individual trials or situations. As people choose to allocate cognitive control based on an increasingly more accurate approximation of the net benefit of exerting it, people gradually become more resource-rational. Unlike strategy selection, the allocation of cognitive control involves choosing a continuous control signal whose intensity determines the likelihood that the response will be produced by the deliberate process rather than the automatic process. This model provided a unifying explanation for a number of phenomena observed in the Stroop task and related experimental paradigms used to study cognitive control (Lieder et al., 2018d), and its novel predictions about the plasticity of cognitive control were confirmed in a new learning paradigm (Bustamante et al., 2021).

So far, we have argued that people can become more resource-rational by learning when to select which cognitive strategy. But what if people encounter an environment that is so novel that none of their strategies is suitable for it? Research suggests that, in this case, people discover novel strategies. Crucially, what strategies people discover in a new environment is not random. Instead, people discover new strategies that are resource-rational for the new environment they are in (Jain et al., 2022). Computational modeling work suggests that the underlying learning mechanism is similar to the rational model of strategy selection learning described above (Krueger et al., 2017; Jain et al., 2019; He and Lieder, 2023). The main difference is that people learn to sequentially select individual cognitive operations rather than an entire cognitive strategy. Critically, this learning mechanism teaches people to choose each cognitive operation exactly in those situations when it is resource-rational, thereby increasing people's resource rationality at a more fine-grained level.

In summary, although strategy selection learning is clearly an important pathway to increased resource rationality, it is not the only one. Two additional learning mechanisms that can make people more resource-rational are i) learning when to override effortless automatic processes by effortful controlled information processing (Lieder et al., 2018d) and ii) discovering new cognitive strategies that are more resource-rational for the current environment (Krueger et al., 2017; Jain et al., 2022, 2019; He and Lieder, 2023).

5.5   Implications for the debate on human rationality and improving human decision-making

The models, experiments, and simulations reviewed in this chapter have two important implications for the debate about human rationality. First, our rational model of strategy selection closes a crucial gap in the adaptive toolbox metaphor of Gigerenzer's theory of ecological rationality (Gigerenzer and Selten, 2002) by specifying how people can efficiently decide when to use which heuristic. Second, our results illustrate one way in which people's capacity for rational action is incrementally refined by learning from experience: Learning from experience teaches people when it is resource-rational to use a given heuristic versus another. Moreover, learning from experience also teaches people new heuristics and when to override their fast automatic processes with slow and deliberate ones. Together, these adaptive learning mechanisms bring human cognition in increasingly closer alignment with the principle of resource rationality.

The experiments presented in this section confirmed the predictions of our resource-rational theory of strategy selection learning. The first experiment showed that people learn to think less when they think too much. The second experiment showed that people learn to think more when they think too little. The third showed that people learn to adapt not only how much they think but also *how* they think to the structure of the environment. The fourth experiment demonstrated that adaptive flexibility also increases with learning, and this enables people's strategy choices to exploit the structure of individual problems. Most importantly, in all four cases, the underlying learning mechanisms made people's strategy choices increasingly more resource-rational.

The empirical evidence presented in this section supports our hypothesis that the human brain is equipped with learning mechanisms that make it more resource-rational over time. Moreover, our subsequent research suggested that learning to become more resource-rational is not confined to strategy selection but extends to the plasticity of cognitive control and the discovery of new cognitive strategies.

Thus, even though people may not be resource-rational when they first enter a new environment, their information processing appears to converge to the rational use of their finite time and bounded computational resources. This perspective replaces the static view that people are either rational or irrational with a dynamic view, according to which people can become less irrational over time by learning from experience. According to this dynamic view, people's capacity to improve their reasoning and decision-making by learning from experience determines how rational they will become.

Overall, our findings paint an optimistic picture of the human mind by highlighting people's capacity for learning to become more resource-rational. This perspective highlights that our rationality is not fixed. Instead, it is malleable and can be improved.

6

Strategy Discovery

In the previous chapter, we saw how people can adaptively select from among a set of strategies for making a decision by balancing outcome quality with cognitive cost. But this leaves open a major question: Where do those strategies come from?

The beginnings of an answer to this question can actually be found in some of the earliest work on effort-accuracy trade-offs in strategy selection, conducted by Payne et al. (1988). They quantified the effort cost of a strategy in terms of the number of **elementary information processes** (EIPs; Chase, 1978) that would be executed when applying the strategy to a particular decision problem. These EIPs included operations like reading a feature value into working memory, adding two numbers, comparing two numbers, and so on. Intuitively, a frugal strategy like Take-The-Best (introduced in the last chapter) requires far fewer of these EIPs than a more accurate strategy like weighted additive. Indeed, Payne and colleagues found that people tended to use strategies that required fewer EIPs as long as they didn't incur a substantial loss in performance.

Beyond decision-making, the idea that complex cognitive processes can be decomposed into simpler cognitive operations goes back at least as far as 1868, to Franciscus Donders's work on "mental chronometry" (Donders, 1969). This idea also played a key role in the earliest computational studies of human cognition conducted by Allen Newell and Herb Simon (Newell and Simon, 1956; Newell et al., 1972), where they decomposed complex tasks such as proving theorems into individual applications of different logical rules. The idea was then taken to its extreme in **cognitive architecture** models, such as ACT-R (Anderson, 1996) and SOAR (Laird et al., 1987), which explicitly model highly specific cognitive operations such as perceptually encoding a stimulus, recalling information from memory, and transforming symbolic representations.

The idea that complex cognitive processes, such as decision-making strategies, are composed of many simpler operations provides a natural answer to

111

the question of where strategies come from. In this view, a strategy is simply a sequence of operations, or more generally—and powerfully—a *rule* for selecting which operation to perform next based on the results of previous operations. Thus, adopting the resource-rational approach, we can reframe the opening question as: What are good (in the cost-benefit sense) rules for sequencing cognitive operations?

Stepping back a bit, if we take seriously the idea that complex cognitive processes can be broken down into sequences of simpler operations, we might even ask: Why bother with specifying and choosing between distinct strategies? Ultimately, the problem facing a resource-rational decision-maker is not one of developing and selecting among strategies. Instead, it is a problem of sequencing cognitive operations in a way that leads to high-utility outcomes while minimizing cognitive cost. Constructing a small set of strategies (i.e., ways of sequencing operations) and selecting from this predefined set for each decision one encounters is *one* way to tackle this problem (and perhaps a good one, as we discuss at the end of this chapter), but it is not the only way. Alternatively, rather than committing to a strategy up front, you could continually adapt the way you're thinking about a problem, identifying unproductive lines of thought, and adjusting priorities accordingly. This more flexible approach may ultimately result in better decisions made with fewer operations, that is, more resource-rational thinking.

Enticing as it may be, this more flexible approach to cognitive strategizing runs into a major challenge. There are an enormous number of ways to sequence cognitive operations. To be precise, there are k^n ways to sequence n operations given a set of k possible operations—and that's not even taking into account different ways the results of early operations could influence the selection of later operations, a critical part of any adaptive cognitive process. How can we possibly formalize, let alone solve, a problem like this? Fortunately for us, we are not the first to encounter this question; we can again draw upon ideas from rational metareasoning. In particular, we adopt the key observation that resource-constrained computation presents a **sequential decision problem**.

6.1 Cognition as a sequential decision problem

A sequential decision problem is a problem that requires making a sequence of interdependent decisions. But what does it mean to say that cognition is a sequential decision problem? One way to understand this idea is as an analogy between the type of problems posed by our external, physical environments and the type of problems posed by our internal, mental environments. To

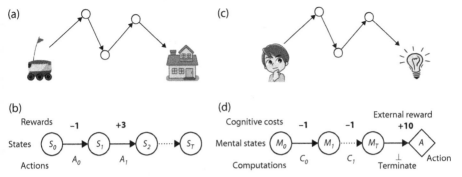

FIGURE 6.1. Sequential decision problems posed by external and internal environments. External environments (a) are often modeled by a Markov decision process (MDP) (b). An MDP is defined by a set of possible states the world can be in, a set of actions the agent can perform, a reward function that encodes the agent's goals, and a transition function describing how actions change the state (arrows). An internal, cognitive environment (c) can be modeled as a *metalevel* MDP (d). A metalevel MDP is defined by a set of mental states and a set of cognitive operations that change that mental state (as specified in the metalevel transition function). The metalevel reward function captures both the cost of executing those operations as well as the external utility of the behavior that the sequence of cognitive operations informs.

make things concrete, consider the problem facing a delivery robot, illustrated in Figure 6.1(a). Completing the delivery will require visiting a sequence of locations before arriving at the final destination. And at each location, the robot will need to decide where to go next. Thus, the robot faces a sequential decision problem. Figure 6.1(b) illustrates how this type of problem is often modeled in artificial intelligence research: a **Markov decision process**, or MDP. At each time step, an agent (here, the robot) takes an **action** (e.g., driving forward), which causes the environment to enter a new **state** (e.g., one where the robot is in a new location). The effect of each action on the state is specified by the **transition function**. After each action, the agent receives a **reward**, a number that captures how good or bad the immediate consequences of the action are. For example, the delivery robot might receive a large positive reward for reaching the destination and a small negative reward every time it moves (capturing the desire to conserve battery life).

MDPs can be understood as the sequential generalization of expected utility theory, as defined in Chapter 3. The critical difference is that we assume the situations the agent encounters are interdependent, with their actions influencing which situation they encounter next. The reward is analogous to utility. Thus, the goal in an MDP is to maximize the total reward received before reaching a terminal state, which marks the end of the task, or **episode**.[1] The solution takes the form of a **policy**, a mapping from states to actions. Adopting the standard notation of s for states, a for actions, T for the transition function, R for the reward function, and π for the policy, we can define the **optimal policy**, π^*, as:

$$\pi^* = \arg\max_{\pi} \; \mathbb{E}_{\substack{s_t \sim T(s_{t-1}, a_{t-1}) \\ a_t \sim \pi(s_t)}} \left[\sum_{t=1}^{N} R(s_t, a_t) \right]. \tag{6.1}$$

That is, the optimal policy for an MDP is the one that maximizes the expected sum of rewards when actions are selected according to that policy (and state transitions follow the transition function). Note that N is the time step at which a terminal state is entered. The MDP also specifies the initial state, s_1, in which the episode begins.

Returning to our delivery-robot example, Figure 6.1(c) illustrates a seemingly very different type of situation: a person trying to come up with a solution to a difficult problem. However, as the diagram suggests, the two cases actually share the same basic structure. Both involve an extended interaction between an agent and an environment; but whereas the robot is interacting with an *external* environment, the problem solver is interacting with an *internal* environment: their own mind. Just as the robot makes several moves, and visits several locations before reaching the destination, the problem solver has several thoughts, and enters several mental states before discovering the solution. Indeed, as illustrated in Figure 6.1(d), this problem can be modeled in precisely the same way as the delivery problem. However, now the actions correspond to cognitive operations and the states correspond to mental states. Thinking changes one's mental state just as moving changes one's physical state. Because this MDP describes the metacognitive problem of how to interact with one's own mind, it is called a **metalevel MDP** (Hay, 2016).

6.1.1   Metalevel MDPs

We provide a formal specification of metalevel MDPs and techniques for solving them in Appendix 10. Here, we provide a broad overview (see Table 6.1 for

---

1. Some formulations instead assume that the MDP runs forever but that future rewards are exponentially discounted. This case can be easily reduced to the episodic case by assuming a fixed probability of transitioning to an absorbing state where all actions yield no reward. However, discounting is often better interpreted as part of the solution than part of the problem, e.g., to capture myopia in cognitive models (Gabaix and Laibson, 2017) or regularize value estimates in AI models (Amit et al., 2020).

Table 6.1. The components of a metalevel MDP.

External Problem/Problem Domain

A **world state** $w \in \mathcal{W}$ specifies the aspects of the external world that are relevant to the agent's current task. We also specify a distribution over possible world states, $p(w)$.

An **action** $a \in \mathcal{A}$ is an external behavior the agent can perform.

The **utility function** U specifies the expected utility of the outcome produced by taking action a in world state w as $U(w, a)$.

Cognitive Architecture

A **mental state** $m \in \mathcal{M}$ captures the architecture's internal state. Often, it will correspond to a belief about the world state, but it can also capture arbitrary variables in a cognitive model. We also specify an initial mental state m_1 with which the agent begins for each new problem.

A **cognitive operation** $c \in \mathcal{C}$ is a primitive operation afforded by the architecture. These operations do not affect the world state, but will typically change the mental state.

The **transition function** T defines the effect of cognitive operations by specifying a distribution of the new mental state given the previous mental state, the executed operation, and the world state. Formally, $m_t \sim T(m_{t-1}, c_{t-1}, w)$.

The **cost function**, "cost," specifies the cognitive opportunity cost associated with executing operation c in mental state m as $\mathrm{cost}(m, c)$.

The **termination operation** $\perp$ indicates that an action should be taken based on the current mental state, ending the episode. All metalevel MDPs include $\perp$ in the set of operations $\mathcal{C}$.

The **action policy**, "act," selects an action to take based on the mental state in which $\perp$ is executed. Denoting that mental state m_N, $a \sim \mathrm{act}(m_N)$.

---

a concise summary). Intuitively, the sequential decision problem of cognition depends on two things: the external problem facing the agent and the internal cognitive architecture with which the agent solves that problem. The external problem and cognitive architecture correspond to the problem domain and domain-specific architecture from Chapter 3. We can thus define the external problem in the same way as we did there, using expected utility theory. Concretely, we define a set of possible **world states** $\mathcal{W}$, a set of **actions** the agent can take $\mathcal{A}$, and a utility function U, such that $U(w, a)$ is the expected utility of taking action a in world state w, marginalizing over possible outcomes.[2] Note that we use w here instead of s to emphasize that it is external to the agent.

To formalize the cognitive architecture, we start with two components that are hopefully also familiar from Chapter 3: a set of **cognitive operations** $\mathcal{C}$, and a **cost function** that captures the opportunity cost associated with executing each operation (note that the $\mathcal{C}$ here has a more narrow definition than in Chapter 3). Now we must depart from familiar ground. In the last chapter, we assumed that the agent engages in one step of metareasoning for each problem, selecting a single heuristic that determines the agent's external action. Our goal now is to increase the granularity of metareasoning, allowing the agent to select multiple (more elementary) cognitive operations before taking an action. This raises a question: If the function of cognition is to guide behavior, how can we quantify the benefit of a cognitive operation that does not lead directly to an external action? Intuitively, cognition can have long-lasting effects because thinking changes our internal representations of the world, and these representations ultimately guide our behavior. More generally, executing a cognitive operation can change the internal state of the agent's cognitive architecture. To formalize this, we augment the cognitive architecture with two new components: a set of **mental states** the agent can hold $\mathcal{M}$, and a **transition function** T, which describes how cognitive operations change that mental state. An inexact, but perhaps useful, way to understand the mental state is as a repository for the incremental progress generated by each cognitive operation.

Having defined the external problem and cognitive architecture separately, we now turn to their integration. How does the agent use their cognitive architecture to solve the problem it is faced with? To answer this question, we can think of the cognitive architecture as implementing the mapping between the world state w and action a. Thus, we need to specify how w enters the architecture and how a comes out. For the former, we assume that w influences the results of executing cognitive operations; formally, w is an input to the transition function: $m_{t+1} \sim T(m_t, c_t, w)$. For the latter, we assume that actions are chosen based on the mental state; formally, we introduce an **action policy**, denoted by "act," such that $\text{act}(m)$ specifies the action taken given mental state m. To ensure that all cognitive cost is captured by the cognitive

---

2. Note that we do not include a transition function here. This reflects the simplifying assumption that the external problem is not itself a sequential problem.

operations, the action policy should be computationally trivial. Additionally, note that w is *not* an input to the act function; this reflects our assumption that the influence of the external world on behavior is fully mediated by cognition.

Having established how cognitive operations and mental states govern the relationship between world state and action, there is one remaining question: When does that action actually happen? Recall from the definition of VOC in Chapter 3 (Equation 3.7) that we introduced a special operation, $\varnothing$, to denote "not thinking at all." In the sequential case (where one may have already done some thinking), this corresponds to cutting off further thinking and taking action. For consistency with the rational metareasoning literature, we call this the **termination operation** and label it $\perp$ (read this as "stop"). Thus, if $\perp$ is executed at time step N (i.e., after $N - 1$ operations have been performed) then the agent will take action $a \sim \text{act}(m_N)$. Intuitively, the choice of when to execute termination operation captures the most fundamental metareasoning problem: How long should you think before taking action? Thus, all metalevel MDPs include $\perp$ in the set of cognitive operations C.

We have now defined all the pieces of a metalevel MDP, which formalizes the sequential decision problem of cognition (see Table 6.1 for a summary). We can now ask: What is the solution to this problem? The solution to a metalevel MDP is a **metalevel policy** π. Like the cognitive policies of Chapter 3, the metalevel policy is a mapping from "states" to cognitive operations. However, for the same reason that we did not allow the agent's external actions to depend on the world state, we assume that cognitive operations are likewise selected based on the mental state alone. Formally, $c_t \sim \pi(m_t)$. We can then define the optimal metalevel policy as:

$$
\pi^* = \arg\max_{\pi \in \Pi} \mathbb{E}_w \left[\mathbb{E}_{\substack{m_t \sim T(m_{t-1}, c_{t-1}, w) \\ c_t \sim \pi(m_t)}} \left[U(w, \text{act}(m_N)) - \sum_{t=1}^{N-1} \text{cost}(m_t, c_t) \right] \right].
$$

$$(6.2)$$

That is, the optimal cognitive mechanism for the environment-brain pair defined by a metalevel MDP is the one that maximizes the expected utility (U) of the action chosen at the end of deliberation ($\text{act}(m_N)$) minus the cost of the cognitive operations (c_t) that produced the mental state upon which the choice was based (m_N). Critically, each cognitive operation is selected based on current mental state ($c_t \sim \pi(m_t)$). As a result, the agent typically cannot immediately think about the most relevant features of a problem; instead, they must use early exploratory thinking to guide more focused future deliberation. Finally, the expectation is taken over all the possible instantiations (w) of the general class of problems defined by the environment. Thus, the mechanism cannot be fine-tuned for one particular situation; instead, it must generalize across a wide set of situations the agent might encounter.

Equation 6.2 represents an elaboration of the resource-rational cognitive policy defined in Equation 3.10. To facilitate comparison, we reproduce a slightly simplified form of that equation here:

$$\pi_{D,C}^{*} = \arg\max_{\pi \in \Pi_C} \mathbb{E}_{s|D} \left[\mathbb{E}_{c \sim \pi(s)} \left[\mathbb{E}_{a|s,c} \left[EU(s, a) \right] - \text{cost}(s, c) \right] \right]. \qquad (6.3)$$

Both equations define an optimal cognitive policy as one that maximizes the expected difference between the expected utility of the action taken and the cognitive cost of the cognitive process that produced that action. Concretely, the objective function involves an expectation over problems the agent might encounter $(s \rightarrow w)$, the cognitive process used to solve that problem $(c \rightarrow (m_{1:N}, c_{1:N}))$, the utility of the resulting action $(\mathbb{E}_{a|c}[EU(s, a)] \rightarrow U(w, \text{act}(m_N)))$, and the opportunity cost of executing that process (a single cost $\rightarrow$ a summation). Note also that the D and C in Equation 6.3 correspond to the external and internal components of the metalevel MDP (Table 6.1). There is just one difference. While Equation 6.3 treats the cognitive process as a single entity, Equation 6.2 explicitly models the cognitive process as a sequence of mental states and basic cognitive operations. On the other hand, Equation 6.2 is a simplification of the general sequential formulation of resource rationality, specified in Equation 3.11. In particular, it is the special case where the world state does not change and only one action can be executed (excluding "do nothing," which is implicitly executed at every time step until the last one). This is certainly a limitation; however, as we show below, many interesting problems have this structure, at least approximately.

Equation 6.2 thus formalizes the sequential metareasoning problem posed by cognition. In doing so, it highlights several key properties of this problem, such as as the important role of mental states in orchestrating individual cognitive operations and the ways in which later operations can be informed by previous ones. It also highlights a potential problem: sequential metareasoning problems are extremely complicated! A naive solution to Equation 6.2 would require enumerating over all possible mappings from mental states to cognitive operations, and evaluating each one by marginalizing over all possible sequences of mental states and operations it might produce before making a choice—a daunting task. Fortunately, a substantial portion of work in artificial intelligence is devoted to solving exactly this type of problem, specifically the fields of reinforcement learning and optimal control theory. By leveraging these general tools for solving MDPs, we can often identify near-optimal metalevel policies using off-the-shelf methods with relatively little custom engineering. Appendix 2 discusses specific techniques for solving metalevel MDPs in more detail.

6.1.2   Contextualizing metalevel MDPs

As illustrated in Figure 6.2, we can understand metalevel MDPs in two different ways. MDP aficionados (feel free to skip this paragraph if this doesn't describe you) can understand metalevel MDPs as a generalization of MDPs to model metareasoning. Concretely, this involves two changes. First, we decompose the state into an internal (m) and external (w) component, requiring that the policy is conditional on only the first component.[3] Second, we partition the actions into one set ($C \setminus \perp$) that only influences the internal state and yields a strictly negative reward (cognitive cost), and a single additional action ($\perp$) whose reward is defined by U and act, and which deterministically transitions to a terminal state (see Section 10.3.5 for a definition of the metalevel reward function).

Alternatively, we can understand metalevel MDPs as a generalization of the value of cognition (VOC) framework introduced in Chapter 3 and applied in Chapter 5. Specifically, metalevel MDPs generalize VOC to the sequential case, where the agent can execute multiple cognitive operations before acting in the world. To make this generalization, we introduced the notion of a mental state, which provides a way for multiple cognitive operations to influence a single action. Note, however, that we can apply the idea of mental state to VOC as well; this is illustrated in Figure 6.2. Specifically, the mental state acts as a mediating variable between the cognitive operation and the action. Although this is equivalent to the original formulation, it highlights how VOC closely corresponds to the final time step of a metalevel MDP. The only difference is that, in the metalevel MDP, the cognitive operation is selected based on the mental state rather than the world state (as in VOC). This is a critical difference, however. In a VOC model, we assume that an agent can decide what to think about based on the true state of the world—or,

3. In this way, metalevel MDPs are similar to POMDPs (Kaelbling et al., 1998), where w corresponds to the latent state and m corresponds to a belief state. All the metalevel MDPs in our applications can be expressed as a POMDP, but this is not true for metalevel MDPs in general. See Section 10.6.4 for further discussion of this point.

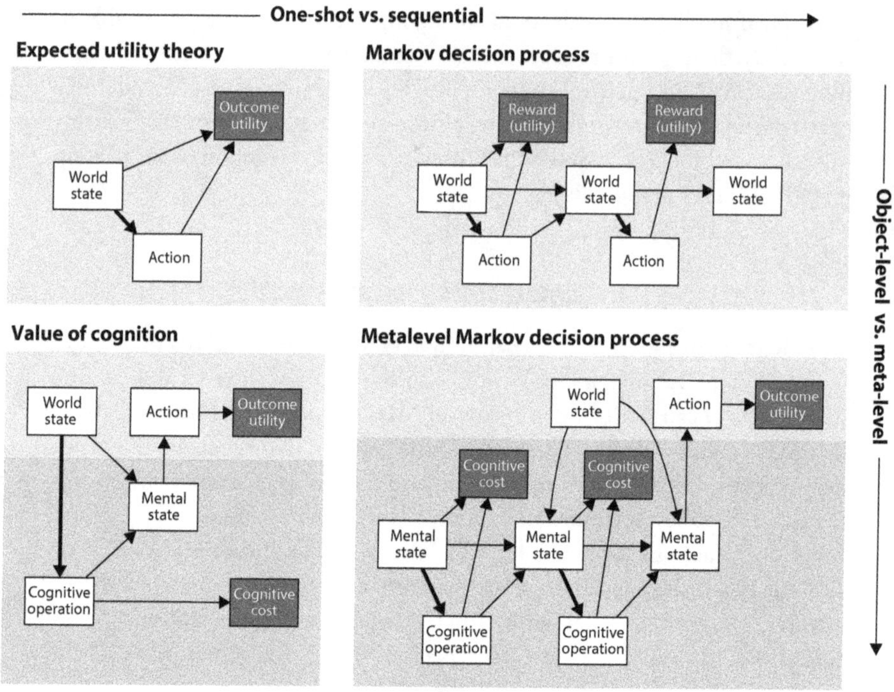

FIGURE 6.2. Understanding the relationship between EUT, MDPs, VOC, and metalevel MDPs. These four frameworks reside in a two-by-two grid defined by two dimensions: one-shot vs. sequential (does the current action influence the next state?) and object-level vs. meta-level (are we optimizing over external actions or internal cognitive operations?). Metalevel MDPs can be understood as a sequential generalization of VOC and/or a metalevel generalization of MDPs. Note that the version of VOC depicted here elaborates on the standard formulation by adding an explicit mental state. Boxes and arrows indicate variables and the causal relationships between them. A bold arrow indicates that the downstream variable is optimized with respect to the upstream variable. Dark boxes indicate optimization targets.

as is usually done in practice, a researcher-specified subset of the world state (e.g., the difficulty of the trial but not the correct response). By contrast, in a metalevel MDP, the agent must decide what to think about next based on what it has already thought about. In this way, metareasoning (or at least part of it, as discussed below) is explicitly modeled as part of the cognitive process. That is, in some cases, the agent will perform a cognitive operation primarily for the purpose of deciding which cognitive operation to perform next.

6.2 Resource-rational analysis with metalevel MDPs

The structure of metalevel MDPs naturally maps onto the five steps of resource-rational analysis, originally defined in Section 3.2. Specifically, steps 1 and 2, in which we specify the problem and class of cognitive mechanisms, correspond to the two main pieces of a metalevel MDP, the external problem and the cognitive architecture (see Table 6.1). We outline the steps in slightly more detail below.

1. *Formalize the problem domain that a cognitive system is optimized for.* We do this in exactly the same way as before, using expected utility theory. That is, we specify a set of possible situations the agent might encounter $\mathcal{W}$, a set of actions the agent can take in response $\mathcal{A}$, and a utility function $U(w, a)$.

2. *Posit a class of cognitive systems that the brain could implement to solve this type of problem, along with the cost of the cognitive resources used by those systems.* In a metalevel MDP, this corresponds to defining the cognitive architecture (the mental states $\mathcal{M}$, the cognitive operations $\mathcal{C}$, the transition function T, the act function, and the cost function) as well as the set of allowable metalevel policies Π.

3. *Identify the cognitive system in this class that maximizes expected utility minus opportunity cost.* In a metalevel MDP, this amounts to identifying the optimal metalevel policy π^*; this can often be accomplished using standard techniques from reinforcement learning.

4. *Derive behavioral predictions from the model and compare those predictions to human behavior.* Deriving predictions in a metalevel MDP amounts to simulating episodes. To simulate an episode, we first sample a world state w and initial mental state m_1. Next, we sample the first cognitive operation: $c_1 \sim \pi(m_1)$. Then we sample the next mental state: $m_2 \sim T(m_1, c_1, w)$. These last two steps repeat until we sample the termination operation, at which point we sample the action: $a \sim \pi(m_N)$. Note that, in contrast to many cognitive models, metalevel MDPs produce very fine-grained predictions about the exact sequence of cognitive operations that underlie a given choice. To test these predictions, we often use **process-tracing paradigms** such as mouse- and eye-tracking. These methods provide rich behavioral data that allow for the strongest test of the model.

5. *Address discrepancies or stop.* To address discrepancies with behavior, we can revise any element of the metalevel MDP or put additional constraints on the possible metalevel policies. Alternatively, if the assumptions are already sufficiently realistic, we stop.

In the remainder of this section, we describe four case studies showing how we have used metalevel MDPs to understand how people adaptively and dynamically allocate their mental resources to guide their behavior while minimizing cognitive cost.

6.2.1   Discovering strategies for risky choice

For our first case study, we return to the problem of risky choice discussed in Chapter 5. To refresh your memory, risky choice involves choosing between multiple actions (e.g., "gambles") whose payoffs depend on a random event (e.g., the color of a ball pulled out of a jar). As reviewed earlier (Section 5.3.2), people adopt a number of different strategies for making these kinds of choices, ranging from accurate but time-consuming strategies that integrate across all the possible outcomes (e.g., WADD) to imperfect but more efficient strategies that only consider the most likely outcome of each gamble (e.g., TTB). We've already seen that people can select between these strategies in a resource-rational way—but why is it *these* particular strategies that people choose between? Are there other strategies that might be even more effective, and, if so, do people use them?

To answer these questions, we designed a metalevel MDP for risky choice. Figure 6.3(a) illustrates the basic ideas of this model. Starting with the external problem, the **world state** defines the payoffs associated with each outcome of each gamble, as well as the probabilities of those outcomes. An **action** corresponds to selecting a gamble. The **utility** function specifies the expected payoff for each gamble.

Turning to the architecture, we assume that a **cognitive operation** corresponds to considering one of the possible payoffs of one gamble and integrating this information into a belief about that gamble's expected utility (specifically, a Gaussian distribution). These beliefs (as well as a record of which payoffs have been considered) are encoded in the **mental state**. We additionally make a simplifying assumption that the mental state also includes the outcome probabilities. The **transition function** specifies how considering a payoff updates the belief about the associated gamble's expected utility; we assume that this update is consistent with Bayesian inference. The **cost function** assigns a fixed cost for each payoff considered. Finally, the **action policy** chooses whichever gamble has maximal expected utility in the current mental state (breaking ties randomly). Recall that this action is taken when the termination operation is executed.

Solving this metalevel MDP yields a resource-rational strategy for risky choice. This strategy corresponds to a policy that chooses which payoff to consider next based on the outcome probabilities and the previously

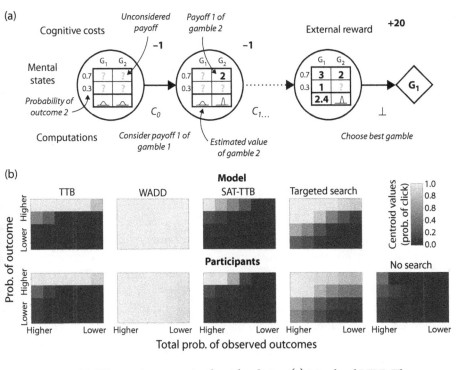

FIGURE 6.3. Discovering strategies for risky choice. (a) Metalevel MDP. The mental state encodes which possible payoffs have been considered and the consequent beliefs about each gamble's expected value. A cognitive operation considers one payoff and updates the agent's belief about the expected value of the corresponding gamble. (b) K-means clustering reveals distinct patterns of consideration, or "strategies," in both simulations from the optimal policy (top row) and information-revealing clicks made by the participants in our experiment (bottom row). The heatmaps show the cluster centroids (i.e., prototypical patterns), where the shade of each cell indicates the probability that the payoff in that cell is considered as a function of the rank of its probability (rows) and the number of previously considered payoffs in that row (columns). We label the clusters by the strategies that they resemble.

considered payoffs. Thus, in contrast to strategies like weighted additive (WADD), which specifies a fixed set of payoffs to consider, the resource-rational strategy can adapt on the fly, for example, focusing on gambles that it has already found to have high payoffs. Furthermore, because the outcome probabilities are part of the state definition, this single policy works well for all kinds of risky choice problems, thus eliminating the need for explicit selection among alternative strategies.

In the previous chapter, we showed how choice problems with different structures call for different decision-making strategies. Thus, in collaboration with our colleague Paul Krueger (Krueger et al., 2024), we applied the policy to a large set of decision problems in which we systematically varied the payoff magnitude, outcome probability dispersion, and opportunity cost (recall that high dispersion means that some outcomes are much more likely than others, encouraging non-compensatory reasoning). Inspecting the optimal policy's behavior, we found several distinct patterns in which payoffs it chose to consider. To quantify this, Paul applied a k-means clustering algorithm to the policy's behavior, finding that a four-cluster solution provided the most parsimonious account. The centroids of these clusters are illustrated in Figure 6.3(b).

Inspecting Figure 6.3(b), we see that the first cluster corresponds closely to the Take-The-Best (TTB) heuristic. That is, the prototypical pattern for cluster 1 is to consider the most likely payoff of each gamble, which corresponds to deciding based on the most informative, or "best," attribute. The second cluster corresponds to the weighted additive strategy (WADD), which considers all possible payoffs of all gambles. This suggests that TTB and WADD are indeed resource-rational strategies for some types of risky choice problems. The next two clusters reflect new strategies, which had not been previously described. Cluster 3 reflects a highly frugal strategy that only inspects payoffs for the most likely outcome (like TTB) and stops when it finds a sufficiently high payoff (like satisficing); we thus refer to this strategy as SAT-TTB. Finally, Cluster 4, which we call Targeted Search, reflects the most sophisticated (and, also, most widely used) strategy. It starts by inspecting some or all of the payoffs for the most probable outcome (as in SAT-TTB), but then inspects additional payoffs for the second-most probable outcome from two or three of the most promising gambles, sometimes continuing to the third-most likely outcome to decide between the top two. This pattern of increasingly focused deliberation highlights the on-the-fly adaptation typical of optimal policies for metalevel MDPs. Importantly, although we have described the behavior of the resource-rational policy in terms of four distinct "strategies," the model does not actually have an explicit notion of strategy; it simply selects individual payoffs to consider, one at a time. The apparent selection among a set of distinct strategies is thus an emergent property of the model, reflecting flexible adaptation to the intrinsic structure of the class of risky choice problems that we considered.

Having characterized the resource-rational policy for risky choice, we next asked if people would exhibit similar behavior. In particular, would they use the same strategies, and in the same situations? To answer this question, we conducted a large-scale online experiment using the Mouselab paradigm introduced in Chapter 5. Each of the 2,368 participants solved twenty problems drawn from one of fifty problem distributions varying along the same dimensions as in our simulations: payoff magnitude, dispersion, and cost (which we operationalized by charging people some amount of money for every click). Applying the clustering method described above to participant's choices of which payoffs to reveal on each trial, we found a strikingly similar set of centroids (Figure 6.3(b)). That is, participants used all the strategies employed by our model—as well as one additional "strategy" corresponding to not considering any payoffs and selecting a gamble at random. Furthermore, across all fifty experimental conditions, we found that participants used these first four strategies about as frequently as the optimal policy did; that is, they rationally adapted their choice of strategy to the types of problems they were presented with. However, participants' strategy selection (and, to a lesser extent, execution) was far from perfect, explaining most of the 39% gap in performance compared to the optimal policy.

6.2.2   Attention in preferential choices

For our next case study, consider the problem faced by a diner at a new restaurant, perusing a long menu. They are presented with a number of options and must select the most desirable one. Going through the entire menu and carefully evaluating each dish is likely to leave our diner with a rumbling stomach and annoyed table mates. Instead, they must quickly scan the menu and focus their attention on the most appealing alternatives. How do we allocate our limited attention when making decisions?

How people make these kinds of preference- or value-based choices is a major area of research in psychology. Although details are still debated, the field has largely converged on the idea that these choices are made by integrating noisy evidence about the value of each alternative, which is sampled over time (Ratcliff and McKoon, 2008; Milosavljevic et al., 2010; Usher and McClelland, 2001). Such **evidence accumulation** models typically assume that information about each option is accumulated in parallel. But intuition suggests that decision-making has a more serial structure, in which we evaluate options one at a time. Furthermore, when the options are presented before us, this attentional bottleneck seems to be tightly coupled to eye gaze, such that we are mostly thinking about whatever option we are currently looking at. Consistent with this idea, longer gaze times bias choice in favor of positive items and against negative items (Armel et al., 2008); this and many subtler links between fixation and choice have been captured by a model that increases the rate of evidence accumulation for the currently fixated item (Krajbich et al.,

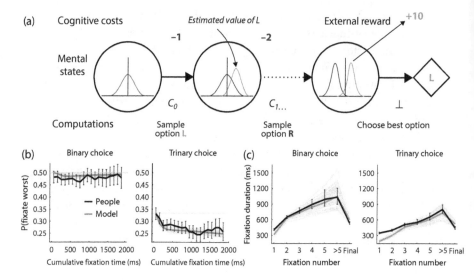

FIGURE 6.4. Attention in preferential choices. (a) Metalevel MDP. The mental state encodes a distribution over the value of each option in the choice set. A cognitive operation corresponds to sampling the value of an option and updating its estimated value by Bayesian inference. (b) Both people and the model (gray) are equally likely to attend to the better or worse option in binary choice (left). In trinary choice (right), however, they are less likely to attend to the worst option in the set as the decision progresses. (c) Both people and the model show increasing fixation durations over the course of a decision.

2010). But if attention drives our evaluation, the natural question becomes: What drives attention?

This problem of attention allocation in preferential choice is naturally cast as a metalevel MDP (Figure 6.4(a)). The **world state** specifies the value associated with having or consuming each item in the choice set. An **action** chooses one of those items, and the **utility function** tautologically states that the utility of choosing an item is equal to the value of having the item. Note that this external problem is a special case of the risky choice problem from the last case study, where there is only one possible outcome.

Turning to the cognitive architecture, the **mental state** represents (Gaussian) beliefs about the value of each item, analogous to the expected-utility beliefs in the previous model. However, we drop the assumption that these beliefs are based on a small, discrete set of features (possible payoffs). Instead, we assume that the agent has access to a continuous stream of information about each option's value (e.g., from episodic memory or mental simulation) but that it can access only one of these streams at a time. Concretely, a **cognitive operation** corresponds to attending to a given option, drawing a single noisy sample of its value, and integrating that information into an estimate of the option's value. The **transition function** specifies that the sample is drawn from a Gaussian distribution centered on the item's true value, along with the consequent Bayesian update of the attended item's value estimate. The **cost function** imposes a fixed cost for each sample drawn as well as a context-switching cost for attending to a different item than the one on the last step (as a technicality, we include that item in the mental state). As before, the **action policy** chooses the item with the highest value estimate.

Before we continue, we should address a possible (and very reasonable) skepticism about the assumption that people represent explicit Bayesian beliefs when making everyday decisions like what to have for lunch. In fact, we make no such assumption. Although we express the model in this way for conceptual and historical reasons (in rational metareasoning, the mental state is called a belief state), we can express an equivalent model in which the mental state simply tracks the total evidence accumulated for each option as well as the number of samples drawn for them (the amount of time spent thinking about each option). Given that the noisiness of the samples is known (an assumption of the model), this representation is isomorphic to a set of Gaussian beliefs (see Section 10.6 for details).

Solving this metalevel MDP yields an optimal policy for allocating attention when making decisions. We compare this policy with human attention allocation using two datasets in which participants chose between junk food snacks (either two or three per trial) while their gaze was recorded with an eye tracker (Krajbich et al., 2010; Krajbich and Rangel, 2011). We can simulate this kind of data from the model by assuming that the attended item is fixated (100 ms for each sample drawn). Callaway et al. (2021) conduct a thorough comparison of the simulated and empirically observed gaze data. Here, we highlight two of the most striking findings.

Figure 6.4(b) shows the probability of fixating on the worst option in the choice set over the course of a decision. When choosing between two items (left), the optimal policy is just about equally likely to fixate on the better item and the worse item throughout the decision. When choosing between three items, however, it is decreasingly likely to look at the worst option as time goes on. Why do we see this pattern? Intuitively, when considering more than two options, one should focus on the top two contenders, as refining the value estimate for these options is most likely to change one's mind about which option is best—indeed, this is exactly what the optimal policy does. When there are only two options, they are necessarily both in the top two, and thus there is no asymmetry. Furthermore, this effect emerges over time because attention must be allocated based on one's *estimates* of value, which only become related to the items' true value after some information has already been sampled.

To highlight a less intuitive prediction, Figure 6.4(c) shows the duration of fixations over the course of a decision. Although the model tends to under-predict the duration of the first two fixations in the three-item case, it captures well three key patterns: (a) the final fixation is shorter, (b) later (but non-final) fixations are longer, and (c) fixations are substantially longer in the two-item case. The first pattern occurs because final fixations are cut off when a choice is made, as in previous evidence accumulation models (Krajbich et al., 2010). The latter two patterns are unique predictions of the optimal model. They arise because more evidence is needed to alter beliefs when their precision is already high; this occurs late in the trial, especially in the two-item case where samples are split between fewer items.

Critically, these qualitative differences in the model predictions in binary and trinary choice are not the result of model fitting. The same set of param-eters are used to generate predictions in each case, and the key qualitative differences are highly robust to the values of those parameters. Put in terms of the previous chapter, two qualitatively different strategies emerge when applied to two variations of the same basic problem, and people move their eyes in a manner consistent with those two strategies.

6.2.3 Thinking ahead

Next, consider the problem faced by a traveler in an unfamiliar country, decid-ing which cities to visit. Geographical concerns limit which cities they can travel between, so their first stop will shape the whole trip. But there are far too many possible routes to consider every one. How do we know which hypothetical courses of action to evaluate, and which to ignore?

How people solve problems that require thinking multiple steps ahead has been a focal question since the very earliest attempts to understand the human mind in computational terms. A key idea from this early work is that thinking ahead can be understood as the construction of a **decision tree**, where edges represent hypothetical future actions and nodes represent hypothetical future states. Initially, this decision tree represents only your current state and the actions you could take there. However, you can expand this tree by imagining taking one of those actions, simulating its outcome (a reward and following state), and adding this information to your decision tree. This operation is called **node expansion**, and is illustrated in Figure 6.5(a). In principle, one can iteratively expand nodes until all possible states have been considered, at

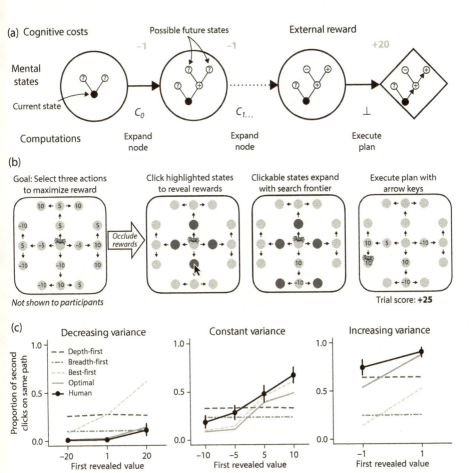

FIGURE 6.5. Thinking ahead. (a) Metalevel MDP. The mental state is a decision tree, which represents possible future actions (edges), states (nodes), and rewards ($+/-$). A cognitive operation corresponds to considering a future state and expanding the decision tree. (b) Experimental task. The internal process of expanding a decision tree is externalized by forcing participants to click to reveal the reward to be gained at each state. (c) Adaptation of planning strategy in environments with different reward structures (see main text). Each panel shows the probability of making a second click on the same path as the first click, depending on the value revealed by that first click.

which point one can quickly determine the optimal plan, that is, the sequence of actions that yield maximal reward.[4]

Most work on human planning has thus focused on identifying the heuristics people use to circumvent the intractability of exhaustive planning. For example, people might limit the depth of their search (MacGregor et al., 2001; Keramati et al., 2016; Krusche et al., 2018; Snider et al., 2015), "prune" away initially unpromising courses of action (Huys et al., 2012, 2015), or avoid planning altogether by relying on habit or "memoization" (Huys et al., 2015; Kool et al., 2017). But such heuristics are generally underspecified: exactly how far into the future should one think, and how bad must an option be to warrant its exclusion from future consideration? We can answer—or even circumvent—such questions by instead asking a more fundamental question. At each moment, we can expand our decision tree in a number of given directions. Which direction should it be?

To answer this question, we model decision tree search as a metalevel MDP (Figure 6.5(a)). Before we start, recall that our definition of metalevel MDPs restricts us to external problems that can be described by expected utility theory. One might naturally conclude that this rules out modeling any form of planning as a metalevel MDP. Surprisingly, this is not the case; but it will require a bit of mental gymnastics. With this in mind, the **world state** in this model is itself an MDP (specifically an acyclic, deterministic MDP). To avoid confusion with the metalevel MDP, we refer to the MDP associated with the external problem as the "object-level" MDP. An **action** thus corresponds to a sequence of object-level actions, that is, a plan. The **utility function** specifies that the utility of a plan is the sum of object-level rewards that will be attained when executing that plan.

The cognitive architecture is a direct instantiation of decision tree search, as described above. Thus, a **mental state** corresponds to a partially constructed decision tree and a **cognitive operation** corresponds to expanding a node. The **transition function** describes the effect of node expansion on the decision tree and the **cost function** assigns a fixed cost for each expansion. Finally,

---

4. Note that decision tree search assumes that the environment is deterministic and acyclic. The former is a significant constraint, but the latter can be easily ensured by adding time to the state. To see how the optimal plan can be quickly determined in a complete decision tree, note that the node expansion operation keeps track of the cumulative reward for all incomplete paths, as well as for the best complete path found so far. Thus, when all nodes have been expanded, the best path is explicitly represented in memory (or it can be reconstructed in linear time by working backward from the optimal terminal node to the root). By also tracking the best incomplete path, the optimal partial plan (if the algorithm is cut off before it completes) can be quickly identified as well.

the **action policy** selects the (partial) plan with maximal cumulative reward under the current decision tree, selecting object-level actions randomly in object-level states that had not yet been considered (see footnote 4 for an explanation of why this does not incur significant cost).

The model described above is deliberately abstract. This is because our concrete implementation of the model makes the major simplifying assumption that the states, actions, and transition function of the object-level MDP are constant and known a priori. The consequence of this assumption is that the agent is implicitly aware of the external environment's graphical structure, and can use this information to guide its choice of operations and actions. Intuitively, imagine running errands in a familiar city: You know which stores exist (states) and how to travel between them (actions and transitions), but you still need to plan a route based on your current shopping needs (rewards).

Solving this metalevel MDP yields an optimal planning algorithm. But how can we compare this algorithm to human planning, given that the latter takes place entirely inside a person's head? To circumvent this challenge, we designed a task that makes people's planning directly observable (Figure 6.5(b)). Inspired by the classic Mouselab paradigm (Payne et al., 1988), our task externalizes planning operations as information-gathering actions. Specifically, participants must click on future states to see what reward they would gain if they visited that state. The sequence of clicks thus reveals the order in which the participant considered each state. This allows us to evaluate candidate models at the level of individual node expansion operations, providing a stronger and more objective test than is possible with previous approaches based on model comparison over the actions people take (e.g., Huys et al., 2015; van Opheusden et al., 2017) or verbal reports of their planning process (e.g., De Groot, 1965; Newell et al., 1972).

In the last chapter, we saw how optimal mental strategies are influenced by the structure of the environment in which they will be applied. To characterize this type of adaptation in the context of planning, we constructed three environments with different reward distributions. In the "constant variance" environment, all states had the same reward distribution. In the other two environments, most states had small rewards and extreme rewards could only be found in one state on each path: the first state in the "decreasing variance" environment and the last state in the "increasing variance" environment. We designed these environments such that the optimal planning algorithm resembled a different classical algorithm in each case: breadth-first for decreasing variance, best-first for constant variance, and depth-first for increasing variance.

Do people likewise adapt their planning strategies to the different reward structures? We can answer this question by looking at patterns in their information-revealing clicks. An especially clear indicator can be found in the second click they make on each trial. After the first click, they essentially have a binary choice: continue exploring the path begun by the first click, or switch to a different path. Figure 6.5(c) shows the probability that our participants continued down the same path as a function of the reward revealed by that first click. In the decreasing-variance case, they tended to switch to a different path, unless they found the rare +20 reward, in which case they usually stop planning and committed to that path; this matches the prediction of breadth-first search. In the increasing-variance case, they tended to keep searching down the same path regardless of the first revealed reward; this matches depth-first search. Finally, in the constant-variance case, they tended to continue searching down paths that began with positive rewards, but switched away from paths that began with negative rewards; this matches the prediction of best-first search. Taken together, we see that participants roughly follow the predictions of each classical search algorithm in the environment in which that algorithm performs well. Only the optimal planning strategy correctly predicts the pattern in all three cases.

So far, we have focused on the cases where people resemble the optimal model. However, we also see systematic deviations from optimal planning. Most notably, when we allowed participants to expand nodes in arbitrary order (not constraining them to forward search), we found a strong bias toward nevertheless considering states in the order in which they would be traversed. This suggests that our metalevel MDP does not capture all the factors constraining human planning in naturalistic settings. In particular, in many cases people may not be able to sample arbitrary future states, and when they can they may have access to a generative model that makes it easier to simulate in temporal order than to reason backward from effect to cause. If people's planning algorithms are adapted to the naturalistic case, we would expect to see discrepancies when these important constraints are removed. This highlights another key strength of resource-rational models: the cases where the model is "wrong" can be just as—if not more—informative than the cases where it's right. In the next chapter, we show how we can use metalevel MDPs to reduce this discrepancy by teaching people more effective planning strategies for environments that don't match their initial expectations or biases.

6.2.4   Feeling of knowing in memory recall

All of our case studies so far have been in the domain of decision-making, a focus that is shared throughout this book. This is a consequence of the core role that expected utility theory plays in metalevel MDPs and resource rationality more broadly. However, metalevel MDPs (and resource-rational analysis) can also be applied to understand other kinds of cognitive functions. Our final case study illustrates this in the domain of memory recall. Specifically, consider the all-too-familiar situation of running into someone whose name you cannot remember. If it feels like you know their name, perhaps even that it's on the "tip of your tongue," you may pause before saying hello. But every moment you delay only exacerbates the awkwardness of the situation. How do people know when to stop searching for a memory?

Most empirical work on metacognition in the domain of memory (or "metamemory") has focused on how people are able to monitor their memory states (Reder and Ritter, 1992; Eakin, 2005) and on evaluating the accuracy of those judgments (Hart, 1965; Vesonder and Voss, 1985; Dunlosky and Lipko, 2007). Less emphasis, however, has been placed on understanding the function of these metacognitive monitoring systems (Schwartz and Metcalfe, 2017). Intuitively, monitoring our memory could help guide how we use our memory, including how long we search for a memory before giving up (Nelson and Narens, 1990). However, a precise computational model of how such a process might unfold has remained elusive.

To fill this gap, we can formalize the decision of when to terminate a memory search as a metalevel MDP (Figure 6.6(a)). Specifically, we consider the case of cued-recall, in which a person is presented with a stimulus (the "cue," e.g., a person's face) and must report an associated piece of information (the "target," e.g., the person's name). Thus, the **world state** corresponds to the target piece of information (the person's true name) and an **action** corresponds to either reporting the target (saying a name) or abstaining (saying "I don't know," or in the name case, "Hey ...buddy!"). The **utility function** assigns a positive reward for correct recall, a large negative reward for incorrect recall, and zero reward for abstaining.

As in the last case study, we make a simplifying assumption to make the problem tractable. Specifically, we abstract away from the concrete information to be recalled and instead model the problem as one of recalling an unspecified piece of information that has a variable degree of strength in memory. Thus, concretely, the world state corresponds to the strength of a person's memory for the target. Admittedly, this is an abuse of terminology, as the strength of a memory is clearly a property of the agent's mind, not the world (a similar objection could be raised for the attention case study, where the world state described a person's preferences). Formally, however, the "world state" in a metalevel MDP simply captures properties of a problem that the agent, more specifically the metalevel policy, does not have direct access to. This formal role is appropriate for memory strength because, as is generally

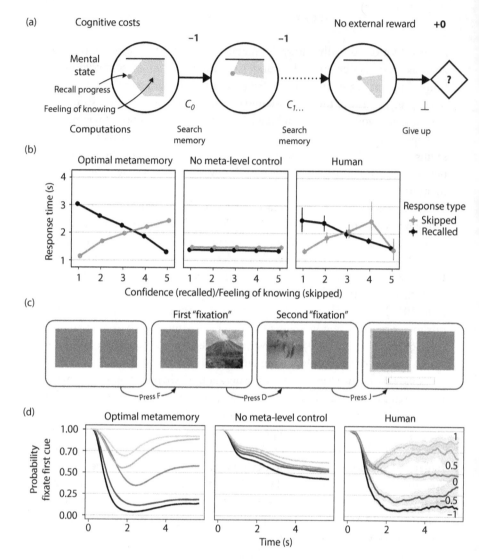

FIGURE 6.6. Feeling of knowing in memory recall. (a) Metalevel MDP. The mental state encodes the current recall progress and an estimate of memory strength (rate of progress). A computation corresponds to searching for the target memory, generating recall progress. (b) Reaction time as a function of metamemory judgment (feeling of knowing for skip trials, confidence for recall trials), separately for trials in which participants correctly recalled the target versus skipped without responding. The models' metamemory judgments are made based on the inferred memory strength at the end of the trial. (c) Multiple-cued recall task. Participants were presented with two images on each trial and could recall the word associated with either of them. Only one cue was visible at a time; participants could flip between them with the D and F keys. At any point they could press J or K to select an image for recall. (d) Time course of attention to the first cue, split by its memory strength relative to the second cue (operationalized by the difference in accuracy in the pretest phase).

assumed in the metamemory literature, people do not have perfect knowledge of the strength of their memories (e.g. Koriat, 1993; Reder, 1987). Note that, under this simplified model, it is impossible to recall an incorrect target. The concrete actions thus correspond to reporting the correct target or abstaining (note that the first action is not always available, as defined by the action policy below).

In this simplified cognitive architecture, a **mental state** captures the amount of progress one has made toward recalling the target memory as well as a metacognitive estimate of the rate of progress (that is, a "feeling of knowing"; Hart, 1965). There is just one **cognitive operation**, which corresponds to continuing to search for the target. Recall that there is always an additional termination operation, which in this case corresponds to giving up on recalling the target memory. The **transition function** describes how recall progress noisily accumulates through search; we assume a Gaussian random walk similar to established diffusion models in memory recognition and recall (Ratcliff, 1978; Sederberg et al., 2008; Yonelinas, 2002). It also specifies how the metacognitive estimate is updated. Specifically, we assume this estimate is a Gaussian distribution over the rate of progress, updated by Bayesian inference as in our previous case studies. Finally, the **action policy** specifies how responses are made when search terminates. In the full model this would correspond to reporting either a guess of the target or the inability to recall it. However, because our reduced-form model does not account for incorrect recall, we instead assume that the action policy deterministically produces the correct response when the recall progress is above a predefined "recall threshold" and abstains otherwise.

Although this reduced-form model is a dramatic simplification of memory recall, which fails to account for some well-established phenomena (most notably, incorrect recall), it may be a reasonable approximation of some recall problems. Intuitively, the model makes three assumptions: (1) a memory explicitly comes to mind only when it reaches a threshold level of subconscious activation, (2) when a memory comes to mind, it can be immediately recognized as correct or incorrect, and (3) the aggregate effect of competing memories on the activation of the target memory is linear and stationary (resulting in a change in the mean and variance of the Gaussian random walk for the target variable). These assumptions may be roughly met in cases where there are a large number of possible targets that are only slightly confusable with each other, for example, recalling a word given an indended meaning.

Solving this metalevel MDP reveals that it is optimal to quickly abandon an attempt to recall a weak memory, that is, one for which recall progresses slowly. The model thus captures the consistent empirical finding that people search longer before giving up on memories for which they report having a higher feeling of knowing (Nelson, 1984; Nhouyvanisvong and Reder, 1998; Gruneberg et al., 1977; Lachman et al., 1979). In one striking demonstration of this phenomenon, Costermans et al. (1992) found opposite relationships between judgments of memory strength and response time when the item was recalled versus not recalled: people gave high confidence judgments for items that were recalled quickly but low feeling-of-knowing judgments for items that were skipped quickly. As shown in Figure 6.6(b), the optimal model reproduces this pattern, while a model without metalevel control (but the same underlying memory recall process) fails to capture either effect.

One weakness with the above finding, and indeed with many empirical results in metamemory, is that we observe at most one single metacognitive decision in each trial, the decision to terminate search. To provide a stronger test of the sequential search-evaluate loop, we designed a modified cued-recall paradigm in which two candidate memories could be recalled on each trial (Figure 6.6(c)). By tracking attention to the corresponding cues using a keypress-contingent display, we could observe this metacognitive process unfolding even before a memory was recalled or given up on. To account for this in the metalevel MDP, we augment the mental state to describe the recall progress and time spent for both possible targets. Similarly, we have two operations, one to search for each of the targets (we also impose a switching cost as in the earlier example). Interestingly, Figure 6.6(d) shows that, contrary to the decision-making case (Figure 6.4(b)), the optimal policy preferentially attends to cues associated with stronger memories even when there are only two options.[5] Participants showed the same pattern. This provides strong evidence that people can estimate the strength of a memory before recalling it and use that information to guide their recall efforts.

6.3   A general framework for characterizing optimal cognitive processes

In this chapter, we proposed that resource-constrained cognition can be understood as a type of sequential decision problem. The key intuition is an analogy between an agent's interaction with the external (physical) environment and its interaction with its own internal (mental) environment (Figure 6.1). By formalizing this intuition with the metalevel MDP framework, we extended the definition of resource rationality presented in

---

5. The intuition for the difference is that, in the recall case, it's not enough to decide which memory is stronger—you actually need to recall a target. This will happen sooner if you attend to the stronger cue.

Chapter 3 (Equation 3.10) to account for the dynamic nature of cognitive processes (Equation 6.2). In our case studies, we showed how this framework can explain and predict how people dynamically allocate their cognitive resources when making choices, planning, and recalling information from memory. Beyond these examples, we have also applied metalevel MDPs in the domains of working memory encoding and maintenance (Ying et al., 2024; cf. O'Reilly and Frank, 2006; Todd et al., 2008; Suchow and Griffiths, 2016) and visual search (Radulescu et al., 2020; cf. Butko and Movellan, 2008; Acharya et al., 2017; Hoppe and Rothkopf, 2019). The breadth of these applications speaks to the generality of the framework. Any cognitive process that can be modeled as a sequence of discrete[6] operations—a large area of cognition, as indicated by the wide application of cognitive architecture models such as ACT-R (Anderson, 1996) and SOAR (Laird et al., 1987)—can also be modeled as a metalevel MDP.

In fact, any model of this form has a corresponding metalevel MDP. Cognitive architecture models almost always specify a space of possible mental states, a set of cognitive operations, and a dynamics function capturing how operations change the state. These correspond to the mental states, cognitive operations, and the transition function of a metalevel MDP, respectively. However, rather than manually specifying explicit rules for which cognitive operation is executed at each time point (e.g., production rules in ACT-R models or decision boundaries in evidence accumulation models), we instead specify a cost function over the cognitive operations and a utility function over the resulting behavior. With these additions, we can use standard tools from AI to identify the optimal policy for selecting cognitive operations. The result is the optimal cognitive process given the constraints of the cognitive architecture (states, operations, and dynamics).

6.4   Addressing the metalevel homunculus

The separation of architecture and policy discussed above is the most powerful feature of metalevel MDPs, but it also presents a significant challenge: identifying optimal (or even near-optimal) policies for metalevel MDPs is often very computationally intensive. In practice, this limits the complexity of cognitive architectures one can consider—although recent technical advances in solving metalevel MDPs using mathematical approximations of

---

6. Note that continuous processes can often be approximated by discrete processes. For example, evidence accumulation models, such as the preferential choice model presented above, are often posed as continuous models but are practically implemented in discrete time.

the value of computation (Hay et al., 2012; Callaway et al., 2018; Sezener et al., 2019; Sezener and Dayan, 2020), hierarchically structured representations of the metalevel policy (Consul et al., 2022a), and deep reinforcement learning (Radulescu et al., 2020; Ying et al., 2024; Chen et al., 2021) suggest possible ways forward. Still, regardless of these technical advances, a corresponding theoretical challenge remains: if solving metalevel MDPs is such a computationally intensive process, how are people able to do it? And, more fundamentally, would it really be resource-rational for them to try? In the cognitive science literature, this problem is referred to as a "homunculus" (Hazy et al., 2006; O'Reilly and Frank, 2006), a "source of unexplained intelligence [necessary] to explain critical features of the phenomena or behaviors in question" (Botvinick and Cohen, 2014, pg. 1256).

To address the first question (how do people find near-optimal policies?), we can model the process by which a person might learn an effective metalevel policy.[7] We took this approach in the previous chapter, and similar models have been applied to the planning task we presented above (Jain et al., 2019; He et al., 2021, 2022; Srinivas et al., 2023; He and Lieder, 2023) as well as to other cognitive tasks (Lieder et al., 2017b, 2018d; Krueger et al., 2017). However, while modeling the learning process handles the "unexplained intelligence" problem (it explains how the homunculus could come to be), we are still left with the second question: Is a controller-architecture separation actually part of a resource-rational solution? There are two ways to answer this second question. One approach is to do away with the separation and embed the controller within the architecture itself. For example, optimal stopping policies in evidence accumulation models can be implemented as time-varying decision boundaries (Drugowitsch et al., 2012), which can themselves be implemented in simple neural circuits (Tajima et al., 2019). A second approach is to embrace the controller-architecture separation and develop computationally tractable models of the controller. For example, O'Reilly and Frank (2006) proposed a model of working memory in which the basal ganglia control the gating of information into the prefrontal cortex. While this does not in itself prove that the controller-architecture separation is bounded optimal, it at least provides a plausible (not to mention, neurally supported) account of how the separation could be efficiently implemented.

Coming full circle, the strategy-selection approach from the Chapter 5 provides another possible solution to the cost of metalevel control. Rather than constantly deciding which cognitive operation to execute next, the controller could instead operate at a more abstract level, selecting a strategy to apply to

---

7. Alternatively, some metalevel policies may be genetically hard-coded and discovered by evolutionary processes.

a given problem, and then letting that strategy unfold with minimal executive guidance. This metalevel strategy of developing and selecting among simpler mental strategies (the "strategy strategy," if you will) could dramatically reduce the cost of metalevel control—as long as the strategies are not too complex and there are not too many of them (a point we return to in Chapter 8).

Importantly however, we are not truly right back where we started. By characterizing resource-constrained cognition as a problem of sequencing cognitive operations, we see that the strategy strategy is just that—a strategy, one possible solution to the more fundamental problem. In fact, the strategy strategy corresponds exactly to a well-studied approach to solving MDPs, **hierarchical reinforcement learning** (HRL; Sutton et al., 1999; Botvinick et al., 2009). In this approach, one seeks to learn a collection of sub-policies for achieving different subgoals, which are strung together by a higher-order policy. Applying HRL to a metalevel MDP, the sub-policies correspond to strategies and the higher-order policy corresponds to the controller that selects which strategy to adopt next. HRL thus provides a tantalizing formal framework for understanding how people jointly learn both the strategies in the toolbox as well as a policy for selecting between them. More generally, modeling cognition as a (metalevel) Markov decision process provides a bridge between reinforcement learning and other cognitive domains. This allows us to import theories about how one should (and how people in fact do) learn how to interact with the external environment to understand how people interact with their internal mental environments.

However, there is one critical difference between mental environments and external environments (at least), as they are typically modeled). While external environments are typically treated as fixed objects that we must adapt to, our internal environments are malleable, and likely subject to the same adaptive pressures that guide our interactions with those environments. That is, we do not just adapt to our mental environments; they also adapt to (or with) us. In the next chapter, we discuss how resource-rational analysis can be applied to understand this kind of adaptation.

7

Representations and Architectures

So far, we have been focused on how resource-rational analysis can be used to identify effective cognitive strategies for solving a problem. In the language of computer science, we have been analyzing the algorithms that intelligent agents should use to solve the problems posed by their environment, making fixed assumptions about the representations those algorithms operate over and the computing architecture in which they are implemented. However, resource-rational analysis can be extended to consider these other aspects of the design of intelligent agents, asking which representations support more efficient algorithms, and what kind of architecture makes it possible to make the most of rational metareasoning.

In this chapter, we consider these extensions of resource-rational analysis, showing how they allow us to engage with classic questions in cognitive psychology. Thinking about representations that support efficient algorithms allows us to explore how people form subgoals and what elements of a complex problem they choose to represent. Asking how architectures should be structured to support effective metareasoning gives us a link to the literature on dual system theories, letting us determine how having multiple different systems that can be used to solve a problem affects performance.

These extensions of resource-rational analysis make direct use of the concepts that we have introduced in the previous chapters. To find effective representations, we assume that people apply a particular algorithm based on those representations and then optimize efficiency over representations rather than algorithms. To evaluate different architectures, we assume that rational metareasoning is used to select between the different systems offered within those architectures, and then optimize over the design of the architectures themselves. In this way, questions about representations and architectures can be reduced to questions about how to efficiently and effectively solve problems while making use of constrained resources—exactly the kinds of questions we have developed tools for answering.

7.1 Resource-rational representation

Anderson (1978) argued that there is a duality between representations and processes: for a given representation-process pair, it is possible to construct a different representation-process pair that produces the same behavior. He made this argument in the context of the classic debate about whether mental imagery used spatial or propositional representations, essentially arguing that this question cannot be resolved by behavioral evidence. However, this argument was also one of the motivations for rational analysis: by abstracting away from the algorithmic level and focusing on computational problems and their ideal solutions, rational analysis provides a way to make progress without having to commit to specific representations and algorithms.

In re-engaging with the algorithmic level, resource-rational analysis needs to deal with the duality between representations and processes. Our general strategy has been to fix a representation and optimize over processes that operate on that representation. Consistent with starting at the computational level, the choice of representation is usually motivated by the definition of a computational problem and is typically relatively innocuous—we assume that people represent things like probabilities, utilities, and states. Where we can, we analyze equivalence classes of processes rather than committing to specific assumptions. For example, our analysis of anchoring-and-adjustment applies to any process that converges geometrically on the correct value, an equivalence class that includes gradient descent and other algorithms in addition to the Markov chain Monte Carlo process we analyze. Similarly, the metalevel MDP models from the last chapter based on evidence accumulation will provide a good approximation for any case where the brain can generate information about a given quantity at a roughly constant rate.

However, the representation-process duality also offers an opportunity to engage with questions where the representation itself is the object of interest. Many questions in cognitive psychology are about how people represent different aspects of the world around them. If instead of committing to a representation we commit to a process, we can use resource-rational analysis to pose a new kind of question about the representation: What representations *should* people use? In the remainder of this section we illustrate how this approach can be applied to two classic questions in cognitive psychology: how people form subgoals, and how people simplify complex environments to plan.

7.1.1 Subgoals

The idea that goals—and subgoals—might be an important aspect of human cognition came to prominence through the work of Newell and Simon on problem-solving (Newell and Simon, 1956; Newell et al., 1972). Being able to identify intermediate steps toward achieving a goal is an important part of planning, and was a key component of the heuristics that Newell and Simon explored for solving challenging problems like theorem proving. More recent work in cognitive science has turned to the question of how people might identify reasonable subgoals for performing a set of tasks. The challenge is thus decomposing those tasks into elements that can be reused to find efficient solutions.

This formulation of the problem of subgoal discovery has a lot in commmon with research in reinforcement learning on how to structure the state and action spaces considered by agents, particularly the idea of discovering "options" that provide agents with high-level actions that target a particular state (Şimşek et al., 2005). Solutions to this problem have tended to focus on the structure of the tasks, constructing graphs in which nodes correspond to states and edges to actions and then analyzing the properties of those graphs. For example "bottleneck" states, which provide a bridge between locally connected regions of the graph, have been proposed as good candidates for such options, and graph-theoretic techniques that partition or "cut" the graph into two pieces provide a way to find them (Şimşek et al., 2005). Such bottlenecks have also been used as a source of hypotheses about how people identify subgoals (Stachenfeld et al., 2017). Other theories of subgoal identification have focused on how people might identify latent causal structure in the environment (Collins and Frank, 2013; Tomov et al., 2020), or find states that support efficient encoding of optimal behaviors (Solway et al., 2014)

Resource-rational analysis provides an alternative to these approaches by focusing on how subgoals improve the efficiency of planning. Correa et al. (2023) proposed a formal framework for making this approach explicit. In this framework, an agent is rewarded based on the quality of a plan it finds, but penalized for the amount of time spent planning. This formulation is extremely similar to that given for planning in Chapter 6, with the objective function containing one term for the value of the plan and another for the cost of planning. However, rather than holding the representation fixed and optimizing the planning strategy, Correa et al. explore the consequences of optimizing the representation over which planning is done, assuming a fixed planning strategy. Specifically, they consider the consequences of including an intermediate state as a subgoal, such that the agent first plans how to reach the subgoal state from the start state and then how to reach the goal state from the subgoal state.

Figure 7.1(a) illustrates the consequences of including specific subgoals for different planning algorithms. Classic planning algorithms such as iterative-deepening depth-first search and breadth first search benefit significantly from the selection of a subgoal that is on the path to the goal.

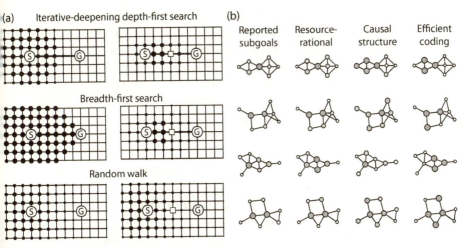

FIGURE 7.1. Resource-rational subgoals (Correa et al., 2023) (a) Effects of introducing subgoals for different search algorithms. Each plot shows a graph where nodes correspond to states and edges indicate actions. The circles labeled S and G indicate the starting and goal states. The size of the black dots reflects the number of times each state is visited. The left column shows the results of simply running the three different planning algorithms, and the right shows the results of introducing a subgoal at the white square, where the algorithms plan to move to the subgoal and then move from the subgoal to the goal. This can result in significant improvements in efficiency as reflected in the number of times the states are visited. (b) A comparison of the subgoals reported by experimental participants with those predicted by the resource-rational model (assuming iterative deepening depth-first search), a model that looks for latent causal structure in the graph (Tomov et al., 2020), and a model that finds efficient encodings of optimal behavior (Solway et al., 2014).

These algorithms involve searching through all states up to a given depth (in this case, the distance from the starting state) before moving on to the next depth. Finding the subgoal allows them to terminate that search and begin a more local search process again starting at the subgoal. Even a random walk can be made more efficient by using a subgoal, as it reinitiates its random exploration of the space at the subgoal and is hence more likely to hit the goal. This makes it possible to pose the question of what subgoals result in the greatest gains in efficiency for a given planning algorithm when we consider an expectation over possible starting points and goals.

Correa et al. showed that the subgoals that maximize efficiency for different search algorithms align with different approaches to subgoal selection that have previously been proposed. Subgoals that are effective for speeding up planning via a random walk over the graph are those with high degree, or that score high on classic measures used for identifying bottlenecks in graphs (Şimşek et al., 2005). Subgoals that help breadth first search are aligned with those selected by methods looking for latent cause structure (Tomov et al., 2020). And subgoals that support iterative deepening depth first search are those that score high on betweenness centrality, a standard option-discovery heuristic that measures how often a state occurs on shortest paths (Şimşek and Barto, 2009). These results help to clarify and organize the existing literature on subgoal selection, showing that the various heuristics that researchers have proposed for identifying subgoals are aligned with increasing efficiency for different search algorithms. They also make it possible to ask where human subgoal selection lies in this space of possible solutions.

Correa et al. conducted a large-scale study in which people learned to perform a number of planning tasks in state-action graphs with different topologies. They then compared the subgoals that people selected as being useful for performing those tasks to the subgoals indicated by existing models and the efficiency-based approach outlined above (four representative graphs are shown in Figure 7.1(b)). They found that people tended to select subgoals that made iterative-deepening depth-first search more efficient; quantitatively, this model provided a better explanation than previous theories of human subgoal choice (Solway et al., 2014; Tomov et al., 2020). Interestingly, betweenness centrality provided a slightly better fit, suggesting that people may have used a simpler heuristic that closely tracks the resource-rational ideal. These results provide a particularly simple account of what seems to drive human subgoal selection.

7.1.2   Construals

Forming subgoals isn't the only way to simplify planning. When faced with a complex environment, another strategy is to use a simplified representation of the environment that captures only the elements that are most relevant to forming a plan. For example, expert chess players quickly identify the elements for board position that are most relevant for selecting their next move. But you don't need to be a chess grandmaster to use this strategy: when making a plan for how we intend to spend our day, we typically abstract away specific details that are unlikely to impact our trajectory.

Finding a simplified construal of the environment to support efficient planning can be framed as a problem of resource-rational analysis. It involves a trade-off between the quality of the resulting plan and the computational cost of representing the environment. This trade-off arises in many settings, but Ho et al. (2022) developed an explicit model of how it might manifest in the context of navigation. They presented people with problems that involved

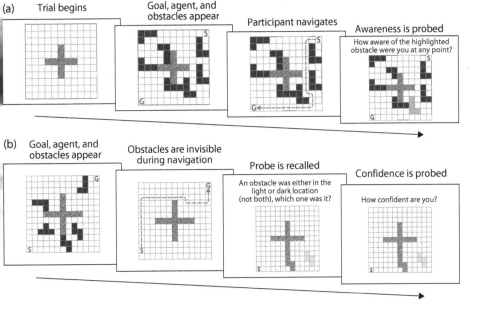

FIGURE 7.2. Two methods of probing people's construal of a navigation problem (Ho et al., 2022). People need to navigate from a starting state (marked S) to a goal state (marked G) while avoiding walls (cross hatched) and obstacles (dark gray). (a) Explicit probing of awareness of obstacles. (b) Exploring representations by testing memory for specific obstacles. Note that the display has been adapted for black and white.

navigating to a goal around a set of obstacles, and then probed the representations people used by either explicitly asking them how aware they were of a particular obstacle or testing their memory for obstacle locations (see Figure 7.2).

This kind of navigation problem can easily be formalized using the now hopefully familiar formalism of Markov decision processes (MDPs). The states correspond to the locations in the grid and actions move up, down, left, and right on that grid. The goal is to get from the starting state to an end state, and achieving that goal is associated with reward. There is a penalty for each action taken, encouraging finding efficient paths. The obstacles manifest via the transition function: if an obstacle occupies a grid location, then the actions that would normally move to that location from other locations will have no effect.

To formalize construals, Ho et al. drew on the key insight that the MDP one uses to plan can differ from the MDP that faithfully defines the navigation problem. For example, in these navigation problems, people might choose to not represent some obstacles, in particular, those that have little impact on planning. To capture this, Ho et al. defined a construal to be a set of obstacles, with the content of this set determining the transition function of the MDP used in planning. Associating each obstacle i with a function $\phi_i(s' \mid s, a)$ that encodes how it constrains movement, the transition function associated with the construal c is given by

$$P_c(s' \mid s, a) \propto \prod_{\phi_i \in c} \phi_i(s' \mid s, a). \tag{7.1}$$

We thus define a unique MDP for each construal, where only the transition function varies across the MDPs.

With this definition of a construal, the problem of forming a construal can be precisely defined. Namely, people should seek to find a construal that allows them to construct plans that result in high utility (in this case, achieving the goal with a small number of actions) while minimizing the number of objects (in this case, obstacles) represented in that construal. This objective function can be expressed in a familiar bipartite form as the utility of the resulting plan minus the cost of representing the construal, which is assumed to increase linearly with the number of obstacles represented. Since this model was intended to identify optimal construals rather than offer an account of how people might form those construals, Ho et al. solved the optimization problem by enumerating possible construals, performing optimal planning for each construal, and then evaluating the resulting utility and cost.

Ho et al. found that people's representations of navigation problems seemed to line up with the optimal construals identified by the definition. Specifically, those obstacles that were unlikely to be included in the optimal construal of a scene were also significantly less likely to be featured in people's representations (see Figure 7.3). Ho et al. complemented these results with a series of further experiments exploring how processes related to memory and execution influenced these construals, and explored how optimal construals might be built on the fly via a more gradual process than the optimization problem we have outlined here.

7.2   Resource-rational architectures

Questions about cognitive architecture are also at the heart of cognitive psychology, with formal treatments of these questions also coming out of the work of Newell and Simon (Newell, 1990). Recently, researchers in several distinct subfields of psychology have explored a particular architectural hypothesis: that the mind consists of two types of systems, one that operates quickly but superficially and another that provides a more careful analysis but

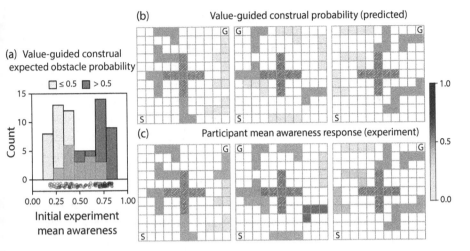

FIGURE 7.3. Results of Ho et al. (2022). (a) Participants reported statistically significantly greater awareness of obstacles that were contained in optimal construals. The light gray indicates overlap. (b) Illustration of the relationship between optimal construals and awareness for three sample navigation problems. In all cases the optimal construal probability (top) aligns well with people's awareness judgments (bottom). Darker shades indicate higher probability/awareness.

does so slowly (Evans, 2008; Kahneman and Frederick, 2002, 2005; Dolan and Dayan, 2013; Diamond, 2013).

Such an architecture might seem to pose a challenge for the resource-rational approach: in the previous chapters, we have considered how people might smoothly vary the resources that they invest. The dual systems approach reduces the number of options one has for such an investment to just two: fast or slow. Why would such an architecture exist? Having more options would support greater flexibility (Payne et al., 1993; Gigerenzer and Selten, 2002). So why would the number of cognitive systems be just two rather than three, 73, or 973? Recent work has explored how human minds seem to choose between the two systems (Daw et al., 2005; Keramati et al., 2011; Lieder and Griffiths, 2017; Shenhav et al., 2013; Boureau et al., 2015). But this still leaves us with the question of why having just two systems might be beneficial.

One way to approach this question within the framework of resource-rational analysis is to consider that there might be some cognitive cost not just to reasoning but also to meta-reasoning. If the cost of deciding which system to use increases with the number of systems available, this might create pressure to have fewer distinct cognitive systems. Milli et al. (2021) explored this hypothesis by determining which set of cognitive systems achieves the best trade-off between the costs and benefits of cognitive flexibility depending on the structure of the environment and the cost of metareasoning. To achieve this, they scored alternative cognitive architectures by the highest possible resource rationality score people would be able to achieve if they used that cognitive architecture. In this case, the resource rationality score is the expected utility of the agent's decision minus the expected cognitive cost of selecting a decision mechanism minus the expected cognitive cost of executing the selected mechanism.

Milli et al. assumed the cognitive cost of selecting a cognitive system is roughly equal to the number of decision systems times the cognitive cost of considering one of them, and referred to the latter as **the cost of metareasoning**. They investigated how the optimal number of cognitive systems depends on the cost of metareasoning and the variability of the environment, which they operationalized by the variance in the importance of the decisions. Across four different types of decision problems, they consistently found the optimal number of cognitive systems increased with the variability of the environment and decreased with the cost of metareasoning.

For many plausible assumptions about the cost of metareasoning and the variability of the environment, the optimal set of decision systems comprised exactly two systems: a fast system that performs no deliberation ("System 1") and a slow system that achieves a higher expected accuracy through deliberation ("System 2"). Figure 7.4 illustrates this finding for the simplest of the four types of decisions: two-alternative forced choice.

The findings of Milli et al. are consistent with a rational reinterpretation of dual-process theories. The human mind might be equipped with exactly two types of cognitive systems because this cognitive architecture achieves the best possible trade-off between the benefits of cognitive flexibility and the cost of choosing and arbitrating between multiple competing cognitive systems. This could have been the case for the environment in which our minds evolved. If so, the analysis by Milli et al. supports the adaptive evolutionary explanation for the architecture of the human mind. Whether our cognitive architecture is still optimal for the environment we live in today is an open empirical question.

7.3   Expanding the scope of resource-rational analysis

In Chapter 1, we stated that we saw resource-rational analysis as an idea that "extends across the psychologist's entire toolbox of cognitive processes." We hope that the material presented in this chapter, together with the results in

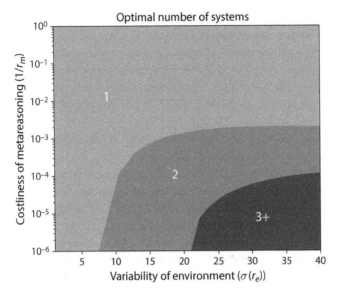

FIGURE 7.4. The optimal number of decision systems for a two-alternative forced choice task increases with the variability of the environment and decreases with the cost of metareasoning. Milli et al. (2021).

the previous chapters, goes some way in illustrating why we hold that belief. Resource-rational analysis provides a way to identify optimal cognitive processes, starting from some assumptions about the problem an agent has to solve, the elementary operations from which a solution is to be built, and the costs associated with those operations. But that framework can be expanded to engage with questions about the representations over which cognitive processes should operate, and even the architecture within which those processes take place. In this way, it provides a theoretical framework in which many of the core questions of cognitive psychology can be expressed.

Consistent with this, there has been a blossoming of research efforts that can be construed as using a resource-rational approach. By now, resource-rational analysis (broadly construed) has been applied to virtually all standard topics of cognitive psychology, including perception (e.g., Wei and Stocker, 2017; Cheyette and Piantadosi, 2020), visual attention (e.g., Hoppe and Rothkopf, 2019; Callaway et al., 2021), working memory (e.g., O'Reilly and Frank, 2006; Suchow and Griffiths, 2016; van den Berg and Ma, 2018), long-term memory (e.g., Lu et al., 2022; Zhang et al., 2023; Callaway et al., 2024), language (e.g., Hahn et al., 2022; Dingemanse, 2020), reasoning (e.g., Dasgupta et al., 2020, 2017; Zhao et al., 2024), problem-solving (e.g., Prystawski et al., 2022; Callaway et al., 2022b; Binz and Schulz, 2023), judgment

(Lieder et al., 2018b), decision-making (e.g., Lieder et al., 2018a; Bhui et al., 2021), active learning (Bramley et al., 2017; Binz and Schulz, 2022; Gong et al., 2023), categorization (Dasgupta and Griffiths, 2022), mental imagery (Hamrick et al., 2015), and moral cognition (Levine et al., 2023). Moreover, resource-rational analysis has also found its way into several other fields, ranging from economics (Gabaix, 2023) to psychiatry (Bari and Gershman, 2024).

This expanded scope also creates other opportunities. The analysis of how multiple cognitive systems might be optimally designed that was presented in the previous section involves two levels of metareasoning: at one level, we assume that people make an optimal choice about which system to use to perform each task; at another, more abstract level, we optimize the design of the architecture within which that choice is made. Allowing multiple levels of metareasoning in this way opens the door to thinking about other ways in which metareasoning could be optimized. For example, perhaps people apply the same kind of task decomposition approach outlined in our analysis of subgoal selection to metareasoning, identifying (features of) mental states that often lead to future success. Adopting this idea may allow us to apply resource-rational analysis to cognitive phenomena such as mind wandering and intrinsically motivated learning that don't immediately ground out in action but instead yield mental states that have unforeseen, long-term benefits. Similarly, just as people can simplify navigation by reconstruing the external environment, people may simplify metareasoning by reconstruing their mental environments, treating certain mental states as inherently rewarding or representing complex sequences of cognitive operations as a single step (the heuristics in Chapter 5 are one example of this).

These kinds of hypotheses can be expressed easily because of the formal framework we have introduced for describing the allocation of cognitive resources as rational metareasoning. By casting the choice of a cognitive strategy as a sequential decision problem and describing that problem in terms of a Markov decision process, we have the opportunity to transfer ideas that have been proposed in the planning and reinforcement learning literatures to the metalevel. Subgoals, options, and the other kinds of abstraction that have been explored in planning and reinforcement learning are equally applicable at this level, and the theoretical literature based on MDPs provides the formal tools for exploring the ideas. We are particularly excited about this as a direction for deepening our understanding of human metareasoning, and for further expanding the scope of our theoretical framework.

8

Improving Decisions

Prior research on judgment and decision-making has identified many systematic errors in human decision-making (Gilovich et al., 2002; Bornstein and Emler, 2001; Kumar and Goyal, 2015; Blumenthal-Barby and Krieger, 2015). These biases cause bad decisions that can have devastating consequences for individuals and society alike.

When looking back on their lives, people often regret the short-sighted decisions they made about their education, careers, and finances (Kinnier and Metha, 1989). The decisions people come to regret were often already irrational at the time. For instance, people's financial decisions often irrationally prioritize their enjoyment of the present over their financial security in retirement (Skinner, 2007), and people's decisions about their health are often similarly short-sighted (e.g., Wang and Sloan, 2018). Such irrational decision-making harms not only the individual but also their family and society at large. For instance, procrastination at work costs the U.S. economy about $10,000 per employee every year (d'Abate and Eddy, 2007).

Moreover, people's biases also seriously compromise their efforts to do good (Caviola et al., 2021). Every year, people in the United States donate 450 billion dollars to charity. Despite this generosity, 700 million people still live in extreme poverty, and many of them die from easily preventable causes. This is mainly because most people's donations go to charities, which are about 100 times less effective at saving and improving lives than the most effective alternatives (Caviola et al., 2021). Unfortunately, the same biases that sabotage people's charitable giving also compromise governmental decisions about public policy and developmental aid (Baron, 1997, 1998). In these decisions, the stakes are even higher, and thus the consequences of errors are even more severe.

Resource-rational analysis helps us understand why people make errors. When considering such explanations, it is important to keep in mind that if devastating errors result from people's reliance on resource-rational heuristics that neither makes them less erroneous nor less devastating. Moreover, although people may often rely on resource-rational heuristics, that's not always the case (Krueger et al., 2024). The errors that result from people's failure to rely on resource-rational heuristics can be much worse. Moreover, in many cases, they can be avoided by using a resource-rational heuristic instead.

In this chapter, we survey the main approaches to improving decision-making and illustrate how resource-rational analysis could, in principle, be applied to make them more effective. Before we begin, we would like to acknowledge that some approaches to "improving" the real-world decisions of individuals, organizations, and society are politically or ethically controversial. Approaches that aim to influence people to take actions they wouldn't otherwise choose are especially controversial because they could infringe on people's autonomy or manipulate them. We believe that interventions designed to influence people's decisions in a particular direction are morally problematic in some cases and morally appropriate in others. The moral appropriateness of such interventions depends on both their goal and their implementation. Whether and how to develop or deploy an intervention to influence people's decisions is a moral decision that should be taken seriously. The risks and moral issues inherent in such interventions should be weighed against its expected benifits. Resource-rational analysis should only be used to develop interventions that are ethical. Our goal in this chapter is merely to illustrate how resource rationality could be applied in principle. The ethicality of specific applications is an important topic for future research.

8.1 Approaches to improving decisions and their limitations

Historically, improving people's decisions has been studied primarily in economics, ethics, law, and public policy. The early approaches focussed on changing the decision problems people encounter so that a rational, self-interested agent will make the right choice. The resulting changes usually take the form of incentivizing good decisions, disincentivizing bad decisions, and restricting the actions people can take. A fourth way to improve people's decisions by modifying the decision situation is to make more beneficial actions possible, for instance, by building infrastructure (e.g., bike lanes), creating new jobs, or supporting technological development (e.g., solar panels).

These traditional approaches aim to give people the opportunities and incentives to positively contribute to society. They generally assume that people's decisions are rational. However, demonstrations of systematic errors in human judgment and decision-making have shown that this assumption is false (e.g., Kahneman and Tversky, 1979; Tversky and Kahneman, 1974). This suggests that the traditional approaches should be reevaluated. The rise of behavioral economics (Thaler, 2015) and behavioral law (Sunstein, 2000) is likely just the beginning of a much more fundamental change in how we foster behavior that benefits society. To update their theoretical foundation, economics, law, public policy, and ethics should consider empirical facts about human decision-making and the best available methods to obtain them. Fortunately, they don't have to start from scratch because researchers in psychology and other social, behavioral, and cognitive sciences have been building this empirical foundation for more than a century.

There is already a substantial body of work on applying insights and methods from psychology and cognitive science to improve human decision-making. A distinctive feature of many of these new approaches is that they strive to improve the cognitive process of decision-making. These cognitive approaches focus on the sequence of thoughts and beliefs through which the decision-maker reaches their eventual decision. Their goal is to improve the fit between people's decision process and the decision problem to be solved. This can be achieved in one of two ways. One can improve what the decision-maker thinks about and how they think about it. Or one can adapt the decision situation to how people make decisions. Approaches that modify the decision situation include restricting choice to good options (e.g., laws prohibiting theft and murder), designing good incentives (e.g., performance bonuses for high-quality work), and nudging (e.g., setting the default choice to the option that is best for most people). Approaches that improve people's decision strategies include providing decision support (e.g., helping someone apply expected utility theory to a real-life decision), debiasing (e.g., teaching people how to avoid the anchoring bias), and boosting (e.g., teaching people when to use the TTB heuristic). We will briefly explain these approaches in the remainder of this section. The subsequent sections of this chapter will lay out how resource-rational analysis can be applied to understand, predict, and improve how these approaches impact people's decisions.

Restricting choice to good options   Many laws improve people's decisions by restricting the options that people can choose with impunity in one of two ways. The first way is to prohibit bad actions, such as stealing and killing. The second way is to mandate that people perform certain actions in specific situations. For instance, drivers are required to stop at red lights and wear seatbelts. Moral rules, social norms, and company policies often serve a similar function (Phillips and Cushman, 2017). When such rules and prohibitions are well designed, they change people's decisions in ways that benefit society overall. The cost of these improvements is that they restrict people's freedom of choice. This cost can be high in exceptional circumstances where violating the rule would lead to better consequences. Moreover, even though many laws, such as the requirement to wear seatbelts, are created to benefit the decision-maker, good intentions don't always lead to good consequences. Moreover, sometimes, people's circumstances, needs, and preferences can radically differ from the situations a law or policy was designed for.

Incentives   The idea behind incentives is simple. If you want someone to take a certain action, you can promise to give them a (financial) reward if they perform it. Alternatively, you can threaten to punish them if they don't. Governments hand out rewards in the form of subsidies, tax credits, and tax deductions, as well as punishments in the form of taxes, fines, and legal repercussions. Companies use financial bonuses to improve their employees' decisions about how much to work and what to do at work. Financial incentives usually motivate workers to invest more effort, but they don't always increase performance and virtually never eliminate irrational decision-making (Camerer and Hogarth, 1999).

When people hear "incentives," they usually think of monetary incentives. However, the concept of incentives is much more general than that. Many action-outcome contingencies can be used to create (dis)incentives. In addition to financial gains, positive outcomes that can serve as incentives include praise, social approval, self-esteem, points scored on a test, attention from parents or peers, social status, prestige, social media "likes" and "shares," awards, grades, pleasant experiences, and future opportunities. In addition to financial losses, negative outcomes that can act as incentives to abstain from an action (disincentives) include criticism, disapproval, social exclusion, loss of trust, respect, or social status, unpleasant experiences, social isolation, guilt or shame, and loss of freedoms and opportunities.

In practice, laws, policies, moral rules, and social norms are implemented via some combination of these nonmonetary incentives, and laws and policies sometimes also impose monetary incentives. We can, therefore, analyze these interventions in terms of incentives.

Nudging   Nudging is one of the more controversial approaches to improving decisions. Nudging influences people's behavior without financial incentives, prohibitions, or penalties. In nudging, a choice architect modifies how the options are presented so that people's intuitive decision mechanisms are most likely to select the option the choice architect deems best. For instance, many governments successfully increased the proportion of organ donors by making organ donation the default option. The citizens in these countries could opt out of being organ donors, but most of them did not. In countries deploying this default nudge, more than 95% of the population are registered organ donors. By contrast, in countries where people have to actively opt in to become organ donors, that percentage is only about 15% (Johnson and Goldstein, 2003). This example illustrates that default nudges can have a large effect on people's decisions without restricting their freedom of choice (but see Jachimowicz et al., 2019; Maier et al., 2022). However, nudging doesn't always work (Jachimowicz et al., 2019; Maier et al., 2022). A recent meta-analysis concluded that, when controlling for publication bias, there is no clear evidence that *the average nudge* is effective (Maier et al., 2022). Some nudges appear to be significantly more effective than others (Jachimowicz et al., 2019). Being able to predict which nudges will be effective would be very useful. Section 8.6 presents a resource-rational framework that can help with that.

Providing decision support   Decision support aims to enable people to make better-informed decisions by performing some of the steps of the decision process for them. This can include information gathering, estimating utilities and probabilities, and calculating expected utilities for them. Historically, decision support was provided by decision analysts—human experts trained in the application of expected utility theory to real-world decisions (Keeney, 1982). Today, decision support is often provided by decision support systems—software that performs the task of a decision analyst on a computer (Marakas, 2003). Decision support systems carry out computations that would be difficult, effortful, or even impossible for people to perform themselves. They thereby provide additional computational resources that augment people's limited cognitive capacities. Below, we will therefore refer to decision support systems as a form of **cognitive augmentation**.

Teaching people better decision strategies   The four approaches presented so far work around the limitations of human rationality. However, these limitations are not immutable. Although some people are more set in their ways than others, the human mind is very malleable. Education, training, and learning from experience can improve how we think.

Teaching the principles of rationality is the most direct way to improve human thinking. This approach is often combined with educating people about the harmful cognitive biases they can overcome by applying those rational principles. For instance, one could point out that people's decisions are overly swayed by extreme events and then teach expected utility theory. The general approach is known as **debiasing** because it aims to help people overcome cognitive biases that presumably impair their ability to make good decisions. The key challenge for debiasing is that applying rational principles, such as expected utility theory, to complex real-world problems is very difficult.

This might be why the benefits of instruction on logic, expected utility theory, and probability theory rarely transfer to the real world (Larrick, 2004).

This problem can be avoided by teaching simple heuristics that exploit the structure of decisions people encounter in everyday life (Gigerenzer and Todd, 1999b; Hafenbrädl et al., 2016). This approach appears to work in the real world, as long as the taught heuristics work sufficiently better than those people already use (Hafenbrädl et al., 2016). This approach has become known as **boosting** (Hertwig and Grüne-Yanoff, 2021). Boosting adapts people's mental strategies to the structure of their environment and their limited cognitive resources. By contrast, nudging adapts the environment to people's heuristics.

8.2 Resource rationality as a theoretical framework for enhancing the rationality of people's decisions

Each of the approaches outlined above can be effective when it is done right. However, each can be implemented in many different ways. Whether a particular implementation will be effective depends on how people will respond to it in the context of the decision problem they are facing. Without a good model of human cognition, this is very difficult to predict. Moreover, to choose the right intervention, we also need to know what constitutes the best decision-making real people can do in the real world. And we also need a good descriptive theory that can predict how people will respond to alternative interventions. Resource-rational analysis is a promising candidate for both.

Resource rationality is not only a promising framework for modeling human cognition. It is also a gold standard against which human cognition can be compared. Recent studies have begun to use the resource rationality score of people's cognitive strategies as a measuring stick. For instance, Krueger et al. (2024) found that while resource rationality can account for a substantial percentage of the suboptimality in human decision-making in a multi-alternative risky choice task, participants could still have performed 38% better at choosing between multiple risky alternatives if they always used the strategy that would be resource-rational for them, given their cognitive constraints. Krueger et al. (2024) found that people underperformed the resource-rational model, partly due to systematic biases in strategy selection. In more complex tasks, the discrepancy between human and resource-rational decision-making is even greater (Consul et al., 2022a; Callaway et al., 2022a; Lieder, 2018a). On the bright side, this suggests that it could be possible to improve human decision-making by teaching people to use resource-rational strategies and giving them computational tools that alleviate their cognitive

FIGURE 8.1. Schematic illustration of the idea that you can use the resource-rational framework to predict which intervention will produce the **maxi**mum improvement in decision-making and then select it. This improvement can be measured by the resource rationality of the resulting decision process (RR). The argmax operator selects the intervention that maximizes this criterion from all available boosts, nudges, cognitive augmentations, and incentives. The resource rationality of the decision process depends on which cognitive policy the decision-maker will use (π), what computational resources the agent's (augmented) cognitive architecture C has available to execute the strategy, and the decision situation (D). Each of the four types of interventions improves one of the three arguments of the resource rationality score RR defined in Equation 3.10. Boosts improve the cognitive policy π. Cognitive augmentation adds cognitive resources to the agent's (extended) cognitive architecture C. Incentives modify the decision situation people are put into (D). Nudging modifies how the decision situation is presented to them (D).

constraints. That is, aligning human decision-making with the prescriptions of resource rationality could be a promising approach to improve human decision-making.

As illustrated in Figure 8.1, resource-rational analysis provides a unifying framework for improving people's decisions through boosting, nudging, incentives, and cognitive augmentation. The key idea is that boosting, nudging, and cognitive augmentation can be thought of as complementary approaches to achieve a common goal: to improve the degree to which human decision-making is resource-rational. Resource rationality depends on multiple factors that can be improved. One of these factors is the cognitive strategies people use, and that's the target of boosting. Another factor is the situations in which people apply those strategies, and that's the target of nudging and incentives. A third factor is the cognitive resources people have available to execute cognitive strategies, and that's the target of cognitive augmentation.

The resource-rational approach to improving decision-making is markedly different from the traditional economic approach. Economists model the action-outcome contingencies outside the decision-maker's head. By contrast, resource-rational models focus on the cognitive processes inside the decision-maker's head. Traditional economics asks, "Under which economic incentives will rational self-interest produce desirable behavior?" By contrast,

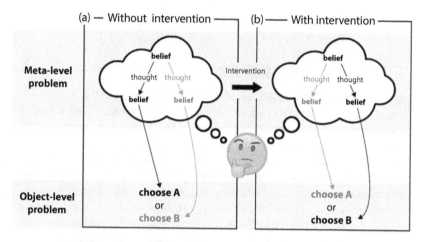

FIGURE 8.2. Rationality enhancement uses resource-rational models of the decision process to design interventions that will, usually, guide the decision-maker's train of thought to a good decision.

the resource-rational approach asks, "How can we make the way in which real people think about their decisions more adaptive for the environment they are in?" To answer this question, resource-rational models specify which piece of information the decision-maker will attend to or recall from memory, depending on what sensory inputs they receive and what they currently believe. They also predict what cognitive operation the decision-maker will perform on that information, how this cognitive operation will change their beliefs, and how that will affect what they will think about next. This is exactly what we need to create interventions for guiding people's trains of thought to the right destination, namely the choice of an action that is objectively good for them and society at large (see Figure 8.2). The decision-maker's train of thought will generally travel through a series of intermediate stops that correspond to, on average, increasingly more accurate beliefs about the relative value of alternative courses of action. Formally, this corresponds to helping people solve the metalevel problem (i.e., deciding how to decide). We call this approach **rationality enhancement** (see Figure 8.2).

What boosting, nudging, and cognitive augmentation have in common is that they try to make people's decision-making "better." The theory of resource rationality provides a precise definition of "better." Moreover, the resource-rational framework also offers three additional advantages over traditional approaches to improving people's decisions because it allows us to i) understand how such interventions affect the decision process, ii) predict the relative effectiveness of alternative interventions and different versions of the same intervention, and iii) derive optimal interventions. This makes it possible to leverage optimization methods to compute optimal boosts, optimal nudges, and optimal cognitive prostheses from realistic models of human decision-making. So far, this idea has only been explored in simple experimental paradigms. Responsibly applying this approach to improve complex real-world decisions of individuals, organizations, and society will likely require decades of concerted interdisciplinary efforts in collaboration with domain experts and stakeholders. Obtaining sufficiently reliable models of important real-world environments will be one of the key challenges. Although the costs of these projects would be substantial, their benefits for society and the future of humanity would be even greater.

The following sections illustrate the resource-rational approach to improving decision-making with demonstrations in simple experimental paradigms. Each of the corresponding resource-rational analyses makes several simplifying assumptions. Many of these assumptions will likely be refined in future applications of resource-rational analysis. Our demonstrations should, therefore, be understood as proof-of-principle illustrations of rationality enhancement rather than as mature interventions based on accurate models.

8.3 Boosting decision-making with resource-rational heuristics

The two existing approaches to teaching rational decision-making reviewed above (i.e., debiasing and boosting) each have limitations. When a debiasing intervention teaches one of the traditional principles of rationality, such as choosing the action with the highest expected utility, students can rarely apply it to the real world. The crux of this problem is that the traditional standards of rationality are unrealistically high. Resource rationality provides a softer standard of rationality that real people can meet in the real world.

How useful it is to boost people's decision-making with a particular heuristic depends, among other things, on how well people perform when they apply that heuristic to real-world decisions. Although we might have already discovered good heuristics for some specific decisions, we are probably still far away from knowing the best heuristics for most types of decisions. This seriously limits how much we can help people by teaching heuristics. For many difficult real-life decisions, the heuristics we can teach people either do not apply or are likely suboptimal. Although using good heuristics is better than using bad heuristics, it would be even better if we could teach people the best possible heuristics. Resource-rational analysis can help us take two steps in this direction: it allows us to define optimal heuristics,[1] and it provides a systematic method to discover them.

This section reviews recent work that has leveraged resource-rational analysis to boost boosting. The foundation of this approach is the definition of optimal boosting.

8.3.1 Optimal boosting

Optimal boosting teaches people heuristics that improve the quality of their decisions in specific types of problems, such as stock market investing and medical diagnosis, for which their default strategies work poorly. We can define each domain as a distinct decision environment (D) with a unique distribution over the possible decision situations s. As a result, the resource-rational heuristic for the environment in which we want to boost people's decision-making will be different from the resource-rational policy for people's everyday lives.

In general, the resource-rational policy π mimics different heuristics depending on which situation it is applied to. However, if the definition of the domain D is narrow enough that all of its decision situations are sufficiently similar, the resource-rational policy becomes equivalent to the best possible heuristic for decisions of that type. When this is the case, we can define the resource-rational heuristic for such decisions as

$$h_D^{\text{boost}} = \arg\max_{h \in \mathcal{H}_C} \mathrm{RR}(h, \mathcal{C}, D), \tag{8.1}$$

where $\mathcal{H}_C$ is the set of heuristics people could use. Recall from Chapter 3 that $\mathrm{RR}(h, \mathcal{C}, D)$ is the expected level of resource rationality that a person with cognitive architecture $\mathcal{C}$ will achieve if they use the decision strategy h in the decision environment D.

If the situations in a domain are more varied, as they are in the medical domain D_{med}, then people should be taught multiple different resource-rational heuristics and when to use which of them. For instance, the resource-rational heuristic for high-stakes medical decisions, such as choosing between cancer treatments, is likely different from the resource-rational heuristic for low-stakes medical decisions, such as choosing between ibuprofen and paracetamol. One should, therefore, derive separate resource-rational heuristics for these different subdomains of medical decision-making. And people should

---

1. The optimality of those heuristics is, of course, subject to corresponding assumptions about the goal to be achieved, the mind's cognitive resources, and the structure of the environment.

be taught a diverse repertoire of such heuristics and how to choose between them (e.g., "For medical decisions about potentially life-threatening conditions, use $h^{boost}_{D_{med, high\text{-}stakes}}$. For decisions about minor medical issues, use $h^{boost}_{D_{med, low\text{-}stakes}}$.").

Optimal boosting comprises three steps. The first step is to select a class of decision problems (D). The second step is to conduct a resource-rational analysis of decision-making in those situations (see Section 3.2). This involves using the computational methods from Chapter 6 to compute a resource-rational heuristic for that class of problems. The third step is to convey this resource-rational heuristic to people so that they will use it in the decision problems identified in Step 1.

The strategy discovery methods from Chapter 6 output raw numbers. This representation of cognitive strategies is not legible to people. This makes it unsuitable for boosting human decision-making. Therefore, to boost human decision-making with resource-rational heuristics, we have to solve one additional challenge: conveying resource-rational heuristics to people. Recent work has demonstrated that this can be done in at least three ways. The first approach generates videos demonstrating what it looks like to apply the optimal strategy (Consul et al., 2022a; Mehta et al., 2022). The second way is to translate the numerical output of strategy discovery methods into a language that people can understand, such as flow charts (Skirzyński et al., 2021) and step-by-step instructions written in natural language (Becker et al., 2022). A third approach is to create cognitive tutors that help people discover, practice, and internalize resource-rational heuristics by giving feedback on people's decision strategies (Callaway et al., 2022a; Heindrich et al., 2025). Let's review each approach in turn.

8.3.2 Teaching resource-rational heuristics by demonstration

Sometimes, the simplest way to convey automatically discovered cognitive strategies to people is to demonstrate them. For instance, the optimal planning strategies we discovered for the process-tracing paradigm illustrated in Figure 6.5(b) are easy to demonstrate because each cognitive operation directly corresponds to clicking on a specific piece of information. This makes it possible to demonstrate each strategy by a video showing the sequence of clicks the strategy performs to make a plan. Showing people these video demonstrations of optimal decision strategies is an effective way to teach them. Several experiments have found that people can learn resource-rational strategies merely by watching such videos (Consul et al., 2022b; Mehta et al., 2022). This approach is not limited to the simple planning task illustrated in Figure 6.5(b). It also works for teaching strategies for choosing between

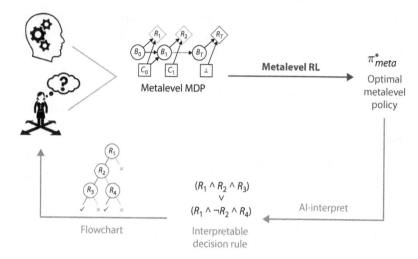

FIGURE 8.3. AI-Interpret translates the output of the automatic strategy discovery methods described in Chapter 6 into human-interpretable flow charts that can be used as decision aids.

multiple alternatives that differ on several attributes (Mehta et al., 2022) and for planning under uncertainty (Consul et al., 2022b). One limitation of this approach is that as the resource-rational strategy becomes more complex, it becomes increasingly difficult for people to understand, remember, and apply it.

8.3.3 Generating decision aids

Understanding demonstrations of an optimal strategy involves inferring rules that explain and predict the strategy's behavior. Skirzyński et al. (2021) developed an algorithm called AI-Interpret that solves this problem for people (see Figure 8.3). Given a series of demonstrations, AI-Interpret infers the series of human-interpretable If-Then rules that provide the most probable explanation of the demonstrated strategy's behavior. AI-Interpret outputs a decision tree for choosing cognitive operations. This decision tree can easily be visualized as a flow chart. The resulting flow charts can be used as decision aids. Skirzyński et al. (2021) showed that providing people with the resulting flow chart significantly improved their performance in three different planning tasks similar to the one shown in Figure 6.5(b). Moreover, they also showed that providing these flow charts was more effective at improving people's performance than providing performance feedback on practice problems.

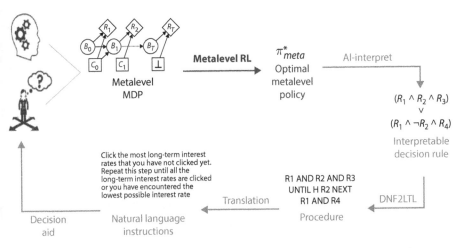

FIGURE 8.4. Translating the output of the automatic strategy discovery methods described in Chapter 6 into human-interpretable procedural instructions in natural language.

Although the flow charts generated by AI-Interpret improved people's decisions, using them was rather cumbersome. These flow charts helped people evaluate whether a cognitive operation they considered performing next was consistent with the optimal strategy. If it wasn't, then people had to test another potential operation, and so on. Following the optimal strategy would be much easier for people if the decision aid told them directly what to do next. To make it possible to create such decision aids, Becker et al. (2022) developed a computational method that translates the output of AI-Interpret into procedural instructions in natural language (see Figure 8.4). For instance, when applied to an experimental task of choosing a mortgage based on its interest rates in the immediate, proximal, and distant futures, this method described the optimal strategy as, "Click the most long-term interest rates that you have not clicked yet. Repeat this step until all the long-term interest rates are clicked or you have encountered the lowest possible interest rate." In a series of two experiments, Becker et al. (2022) showed that the resulting decision aids significantly improved participants' performance in the mortgage task as well as in a planning task. Moreover, they found that such procedural instructions are more helpful than the flow charts produced by AI-Interpret.

So far, this approach has only been applied to artificial laboratory tasks. Scaling it up to real-life decisions, such as medical diagnosis and treatment planning, investment decisions, purchasing decisions, and hiring decisions, is an interesting direction for future research.

8.3.4   Cognitive tutors and cognitive training

While providing people with a decision aid for a specific decision would likely improve how they make that decision, this could be the only decision that will be improved. To achieve broader and more long-lasting improvements in decision-making, it could be more helpful to improve people's decision-making skills by teaching them resource-rational decision strategies. Learning new skills usually requires practice, and decision-making skills are probably no different.

Practice is more effective when it provides immediate, accurate feedback (Kahneman and Klein, 2009). Such high-quality feedback is rarely available when people practice cognitive skills for several reasons. First, it is difficult for teachers to provide feedback on their students' thoughts because they cannot observe them. Second, the effect of unobservable thoughts on observable outcomes is often indirect, delayed, and stochastic. Third, many decisions don't arise from just a single thought. Instead, they are produced by a long sequence of thoughts. This makes it unclear how much any one of those thoughts contributed to the final decision. To overcome these problems, we developed intelligent cognitive tutors that give people immediate, accurate feedback on individual cognitive operations as they practice decision-making (Callaway et al., 2022a).

As illustrated in Figure 8.5, our intelligent cognitive tutors teach resource-rational decision strategies by giving people feedback on each cognitive operation they perform in a process-tracing paradigm such as the Mouselab-MDP paradigm shown in Figure 6.5(b). This **metacognitive feedback** is positive when the participant follows the resource-rational strategy and negative when they deviate from it. Worse cognitive operations receive larger penalties. Moreover, when the participant chooses a suboptimal cognitive operation, the cognitive tutor informs them what the optimal strategy would have done instead. Figure 8.6 illustrates the tutor's metacognitive feedback on a suboptimal planning operation. In a series of experiments, we consistently found that our tutors' metacognitive feedback significantly increased the proportion of participants who discovered the optimal strategy compared to people who practiced without feedback and participants who received conventional feedback on their final decision (see Figure 8.7). Moreover, the effect of practicing planning on the simple task illustrated in Figure 6.5(b) transferred to a more complex and more naturalistic planning task in which participants planned a road trip by using a search engine to look up the costs of hotels in different cities (Callaway et al., 2022a).

While the initial cognitive tutors were limited to small and unrealistically simple planning problems, subsequent work has made discovering and

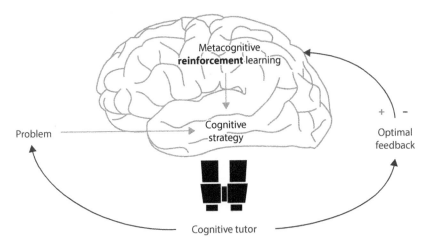

FIGURE 8.5. Illustration of the metacognitive feedback mechanism intelligent cognitive tutors used to teach resource-rational decision strategies. Metacognitive reinforcement learning is a learning mechanism that optimizes how people think based on its consequences.

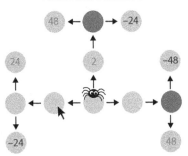

FIGURE 8.6. An intelligent cognitive tutor teaching a resource-rational planning strategy using metacognitive feedback.

teaching resource-rational planning strategies more scalable (Consul et al., 2022b). Concretely, Consul et al. (2022a) extended intelligent cognitive tutors to larger and more complex planning problems. In addition, Hein-drich et al. (2025) extended intelligent cognitive tutors to a more realistic

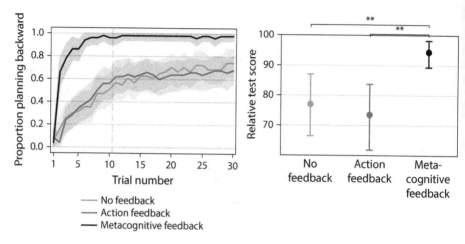

FIGURE 8.7. Effectiveness of a cognitive tutor teaching a resource-rational planning strategy ("planning backward") on learning during practice (left) and performance after practice (right). The results are from Experiment 1 in Callaway et al. (2022a).

planning model, according to which people have to think about each action multiple times to gradually reduce their uncertainty about its outcomes. Most recently, Heindrich and Lieder (2025) developed an intelligent tutor teaching an optimal decision strategy for project selection. Moreover, while the initial approach to developing intelligent cognitive tutors assumed an accurate model of the decision environment, subsequent work made automatic strategy discovery and intelligent cognitive tutors robust to potential errors in the model of the decision environment (Mehta et al., 2022). While these approaches train decision-making in simulated environments, the optimal metacognitive feedback method can also be applied to training decision-making and cognitive control in the real world (Wirzberger et al., 2024).

Providing feedback is not the only way to help people discover resource-rational decision strategies. Another approach is to help people learn from their experience by systematically reflecting on it (Becker et al., 2023). Becker et al. (2023) implemented this approach to boosting by asking people a series of Socratic questions about how they made their decision; why it turned out the way it did; whether, and, if so, how, they could have reached a better decision; and how they planned to apply their lessons to future decisions. They evaluated this intervention in an experiment in which participants made a series of decisions in the same planning task. Participants who were asked the Socratic questions were significantly more likely to discover the resource-rational planning strategy than participants who practiced the task without being prompted to reflect. Prompting systematic reflection by asking Socratic questions is a very general approach that can be applied to boost learning from real-life decisions. Formulating such questions doesn't require knowledge of the resource-rational strategy. Moreover, a single set of Socratic questions might be sufficient to boost a wide range of real-world decisions in different contexts. This makes fostering systematic reflection a promising approach to improving the rationality of human decision-making.

8.3.5 Summary and future directions

The theory of resource-rational boosting outlined in this section provides a precise and principled normative foundation for increasing the rationality of human decision-making. This makes it a promising theoretical framework for understanding and improving the effectiveness of boosting. The findings reviewed in this section suggest that grounding the design of boosts in resource-rational analysis can help us overcome the impasses of the field's previous approaches to debiasing (Larrick, 2004).

We hope that future work will apply the methods reviewed in this chapter to improve the effectiveness of ongoing efforts to enhance the rationality of decision-making in the real world. Many efforts to boost people's decisions by teaching them clever heuristics are already underway in several real-world contexts, including medicine and finance (Hertwig and Grüne-Yanoff, 2021; Hafenbrädl et al., 2016; Gigerenzer and Todd, 1999b; O'Connor et al., 1999). Most applications of boosting don't use computational methods yet, but some of them already do (e.g., Phillips et al., 2017). For instance, Phillips et al. (2017) developed a toolbox for generating decision trees that achieve a good balance between decision quality and effort (Phillips et al., 2017).

The resource-rational theory of boosting introduced in this chapter provides a principled way to make the numerous design decisions involved in designing boosts. As a first step, resource-rational analysis could be applied to analyze and optimize the design of fast-and-frugal heuristics that are currently taught and recommended. At first, the framework can be used to optimize parameters within the set of heuristics proposed previously. For instance, resource rationality could be applied to derive optimal fast-and-frugal trees. This might lead to improvements in domains that warrant more deliberation than developers of fast-and-frugal trees usually recommend. Later on, optimal boosting could be applied to improve upon the best fast-and-frugal heuristics by applying the methods for discovering resource-rational heuristics to a broader class of potential strategies.

Moreover, the theory of resource rationality could be used to determine which approach to boosting is most appropriate for a given domain. For instance, the resource-rational theory of strategy selection introduced in Chapter 5 could be used to predict when a fast-and-frugal tree would be most beneficial and when a different kind of boost might work better. Moreover, the automatic strategy discovery methods introduced in Chapter 6 make it possible to leverage AI to derive heuristics that psychologists have not discovered yet and might never discover on their own.

The breadth and flexibility of the resource-rational approach to boosting could be a crucial advantage over the more narrow approaches that have grown out of the literature on fast-and-frugal heuristics (Gigerenzer and Todd, 1999b). Although fast-and-frugal heuristics have been successfully applied to improve decision-making in specific domains (Hafenbrädl et al., 2016), they are less helpful in other domains. Most fast-and-frugal heuristics were originally designed for categorization decisions. However, many important real-life decisions involve planning, problem-solving, and optimization. Shoehorning such decisions into the classification framework may be suboptimal. The resource-rational framework is more general and can be applied to derive optimal strategies for many additional domains.

Optimal boosting is *not* constrained to decision-making, either. In principle, it can be applied to discover and teach optimal cognitive strategies for any mental function. This includes problem-solving, deductive reasoning, reasoning under uncertainty, mental arithmetic, categorization, and inductive learning. For instance, resource-rational boosting could be used to develop cognitive tutors for problem-solving or reasoning under uncertainty. Intelligent Tutoring Systems are already widely used in education and about as effective at teaching algebra as human teachers (VanLehn, 2011; Koedinger et al., 1997; Kulik and Fletcher, 2016). Many of these tutors model the cognitive skills to be taught using cognitive architectures (Nkambou et al., 2010), such as ACT-R (Anderson, 2013). Resource-rational analysis could be applied to automatically derive the optimal cognitive strategies students should be taught. As far as we know, this approach has not been explored yet. Instead, formulating the strategy an intelligent tutor teaches still relies almost exclusively on human intuition. Computational methods for discovering resource-rational strategies could be used to assist educational designers in discovering even better strategies. Crucially, our methods can also be used to create intelligent tutoring systems teaching optimal cognitive strategies for general cognitive skills that none of the existing intelligent tutoring systems can teach yet.

Moreover, recent advances in large-language models (Zhao et al., 2023; Minaee et al., 2024) could be leveraged to improve and scale up our approach to translating automatically discovered resource-rational cognitive strategies into natural language. This makes it possible to rapidly generalize our approach to creating decision aids (Becker et al., 2022) to many different types of decisions in various real-world settings. Moreover, once our methods have been applied to discover cognitive strategies for other cognitive skills, such as reasoning and problem-solving, large language models can be used to explain them to people. The rapid progress that the fields of decision support and intelligent tutoring systems have experienced since the introduction of large language models suggests that this approach could be very promising (Nye et al., 2023; Vavekanand et al., 2024; Liu et al., 2023; Kasneci et al., 2023; Thirunavukarasu et al., 2023). Therefore, in the long run, applications of improved versions of our methods of resource-rational analysis and automatic strategy discovery could potentially lead to breakthroughs in decision support and education.

8.4   Cognitive augmentation

Boosting helps people make the best possible use of their cognitive resources. But even with the best possible boosts, people's resources will still be limited. Many real-world problems are so complex that people's limited resources are simply not enough to reliably compute the optimal solution, no matter how cleverly those resources are used (van Rooij, 2008). For instance, the complex planning and decision problems people have to solve to achieve their goals are often prohibitively difficult (Reichman et al., 2023). So, instead of extensively planning the simultaneous pursuit of all of their goals using a precise mental model, people often rather apply simple, relatively short-sighted heuristics (Prystawski et al., 2022; Huys et al., 2015) to a simplified model of one sub-problem at a time (Ho et al., 2023, 2022; Correa et al., 2023). These heuristics are thought to be resource-rational. Yet, there are, nevertheless, many real-world situations in which they lead people astray. If these problems result from people's use of resource-rational heuristics, then optimal boosting can't help because teaching resource-rational heuristics is the best that boosting can do for people. In that case, the real bottleneck is people's limited cognitive resources. One way to overcome this bottleneck is to provide people with additional computational resources. This approach is known as **cognitive augmentation**.

Cognitive augmentation is already commonplace. We help people transcend their limited memory by giving them tools for storing information outside their heads, such as calendars that remind them of their appointments, and information retrieval technologies, such as Google Search. Calculators and computers help people transcend their limited computational capacities. Navigation systems, such as Google Maps, enhance people's spatial skills. Smart glasses can enhance not only people's perceptual abilities but even their social cognition (Rashidan et al., 2021; Poglitsch et al., 2024). Finally, people increasingly rely on large-language models, such as ChatGPT, to enhance their cognitive abilities in numerous other ways, and many more AI assistants might emerge in the future (Bahrini et al., 2023).

You can think of these technologies as **cognitive prostheses** because they augment people's cognitive abilities. Given that we can develop different types of cognitive prostheses, which ones should we develop first, and how should we design them? Resource-rational analysis can inform these decisions by identifying cognitive limitations that are a critical bottleneck to people's ability to solve problems crucial to their personal lives and to the functioning of society. In the remainder of this section, we outline how resource-rational analysis can help innovators identify promising opportunities for developing cognitive prostheses and summarize some preliminary steps we have taken in that direction. We start by defining optimal cognitive augmentation as providing those cognitive prostheses that increase people's resource rationality the most.

Each cognitive prosthesis provides its users with some additional computational resource c, such as a calculator or a navigation app. This augments the cognitive resources of the user's cognitive architecture from $\mathcal{C}$ to $\mathcal{C} \oplus c$. This increases the set of possible cognitive strategies from $\Pi_\mathcal{C}$ to $\Pi_{\mathcal{C} \oplus c}$, potentially enabling the user of the cognitive prosthesis to use more accurate cognitive strategies. In addition, cognitive augmentation may increase the accuracy of existing cognitive strategies or decrease the cost of executing them, thereby increasing a fixed strategy's resource rationality from $\mathrm{RR}(\pi, \mathcal{C}, D)$ to $\mathrm{RR}(\pi, \mathcal{C} \oplus c, D)$. For instance, providing a calculator makes strategies that involve calculating expected values more resource-rational by making them more accurate and less costly.

The computational resource whose addition confers the greatest benefit is the one that maximizes the resource rationality of the best strategy that the augmented person can use. Providing that external computational resource ($c^\star$) from the set of all available options ($\mathcal{C}_{\mathrm{ext}}$) constitutes optimal cognitive augmentation as defined in Equation 8.2. This equation illustrates that optimal cognitive augmentation depends on what strategy the augmented person will use and how well it will work:

$$c^\star = \arg\max_{c \in \mathcal{C}_{\mathrm{ext}}} \ \arg\max_{\pi \in \Pi_{\mathcal{C} \oplus c}} \mathrm{RR}(\pi, \mathcal{C} \oplus c, D). \qquad (8.2)$$

To illustrate this abstract idea, we now turn to a concrete example of how cognitive augmentation could be used to help people solve complex problems. When people face problems whose size and complexity exceed their cognitive capacity, they break them down into smaller and more manageable problems by setting subgoals (Newell et al., 1972). With the right subgoals, people can solve, with much less cognitive effort, problems that would otherwise be too difficult (Correa et al., 2023; Maisto et al., 2015). In other words, having the right subgoals can substantially increase the resource rationality of human problem-solving.

Unfortunately, unless the problem has some obvious structure, setting the right subgoals can be an extremely difficult problem. Therefore, a cognitive prosthesis that computes subgoals whose pursuit allows people to efficiently solve complex problems would score very high on the criterion of optimal cognitive augmentation (Equation 8.2). Singhi et al. (2023) developed a cognitive prosthesis that does just that. Given a model of the problem to be solved and a model of human goal pursuit, this cognitive prosthesis uses a global optimization algorithm running on a supercomputer to compute the subgoal that maximizes the predicted resource rationality of people's efforts to solve the original problem.

Singhi et al. (2023) evaluated the effectiveness of this cognitive prosthesis in a dynamic optimization problem that simulated the challenges involved in managing an ecosystem with multiple competing goals: maximizing the yields of several crops simultaneously, controlling pests, and preventing overgrowth that can cause the ecosystem to collapse. They found that the subgoal computed by the cognitive prosthesis (e.g., to eliminate the space worm pest first) significantly increased the proportion of people who performed well on the task. Moreover, participants supported by the cognitive prosthesis deployed lower amounts of pesticides and fertilizers. Although participants were only provided a single subgoal, the method can also compute sequences of subgoals/milestones that pave the path to the final goal.

The findings from this experiment suggest that a cognitive prosthesis that computes subgoals for people could, in principle, help them achieve larger goals and solve complex problems. However, these findings were obtained in a game-like simulated environment. Whether this approach can also work in the real world remains to be seen. An initial attempt to support people with AI-generated subgoals in the real world backfired, presumably because people's estimates of the values of their long-term goals and how long it would take to achieve them were biased and inaccurate (Lieder et al., 2022a). Developing a cognitive prosthesis that helps people set better goals in the real world thus remains a challenge for future work. Future work on this and other applications of cognitive augmentation will need to address the challenges that arise from people's limited information and inaccurate beliefs about the world.

8.5 Optimal incentive structures for boundedly rational decision-makers

Designing (financial) incentives for improving people's decisions is central to many goals of economics. Economic theories of optimal incentive design generally assume that people's decisions are guided by rational self-interest (Bolton and Dewatripont, 2004; Laffont and Martimort, 2002; Börgers, 2015). Not only is this assumption clearly false (Kahneman and Tversky, 1979; Thaler, 2015), but it is systematically wrong in ways that can make the resulting incentives backfire and cause us to overlook highly cost-effective ways to significantly improve people's decisions (Fehr and Falk, 2002). Research in behavioral economics has therefore called for more realistic models of human decision-making that take people's limited rationality and concern for others into account (e.g., Thaler, 2000; Camerer and Fehr, 2006). Subsequent research developed resource-rational models that address these concerns (e.g., Gabaix, 2023, 2019, 2020) and applied them to design optimal incentive structures (Farhi and Gabaix, 2020; Lieder et al., 2019).

Which monetary or nonmonetary incentives are most likely to achieve a desired effect depends on how people respond to the incentives. Choosing good incentives, therefore, requires models of human decision-making that can predict the effects of incentives with sufficient accuracy. Switching to more realistic models, such as resource-rational models, can make these predictions more accurate. This, in turn, makes it more likely that the chosen incentives will be highly cost-effective.

Resource-rational analysis can inform the design of optimal incentive structures in two ways. First, the optimal incentives i^* for achieving a goal g depend on what the resource-rational policy $(\pi^*_{D,C})$ will choose when the person encounters an incentivized situation $(s \oplus i)$ in the modified decision environment $D \oplus i$ (Equation 8.3). Second, the optimal incentives may change what the resource-rational heuristic is in the first place because they change the environment from D to $D \oplus i$ (Equation 8.4). If so, people will likely adapt their strategy accordingly (Krueger et al., 2024; Lieder et al., 2018c; Lieder and Griffiths, 2017). For instance, when the stakes are higher, people think harder (Lieder et al., 2018c) and may consider some information they would otherwise ignore (Krueger et al., 2024). Critically, the predictions of this resource-rational theory of the effects of incentives may be substantially different—and more accurate than—the predictions of the homo economicus model.

$$i^* = \arg\max_{i \in \mathcal{I}} \underset{s|D}{\mathbb{E}} \left[g(\pi^*_{D \oplus i, \mathcal{C}}, \mathcal{C}, s \oplus i) \right], \tag{8.3}$$

$$\pi^*_{D \oplus i, \mathcal{C}} = \arg\max_{\pi \in \Pi_\mathcal{C}} RR(\pi, \mathcal{C}, D \oplus i). \tag{8.4}$$

In some applications, such as encouraging people to save for retirement, encouraging people to invest in their health, and motivating students to learn, the purpose of the incentives is to help people do what is best for them. The optimal incentives for interventions such as these are defined by the simple formula in Equation 8.5:

$$i^* = \arg\max_{i \in \mathcal{I}} \arg\max_{\pi \in \Pi_\mathcal{C}} RR(\pi, \mathcal{C}, D \oplus i). \tag{8.5}$$

Let's now turn to a concrete example of how optimal incentive design can be used to motivate people to make progress towards their goals. When a person's goals are distant, discovering a sequence of actions that will achieve them may require planning multiple steps ahead. In situations like that, people who don't plan ahead underestimate the value of taking a step forward because they only see its immediate costs but not its long-term benefits. As a result, they often put off taking any steps toward their goal until it is too late (Klingsieck, 2013). As a result, people often fail to pursue goals that a rational agent would jump on. The standard homo economicus model does not predict such irrational decisions, making it an unsuitable foundation for deriving incentives that help people overcome this problem. A resource-rational model of human planning would be more suitable (Callaway et al., 2022b). A key feature of such models is that they plan only a limited number of steps ahead and then choose the action that maximizes the reward that the agent foresees accruing within the planned period. These resource-rational models, therefore, predict that when none of the considered plans achieves the goal and all steps towards the goal are unpleasant, then the goal won't be pursued.

According to this resource-rational analysis of planning, the optimal incentive structure (Equation 8.5) would align each action's immediate reward with its long-term value. Incentives creating such an environment would enable people to achieve the highest possible resource rationality score by looking only a single step ahead, thereby gaining the highest possible utility at a minimal cognitive cost. Lieder et al. (2019) showed that such an environment can be created by computing the incentive $i(a; s)$ for taking action a in state s according to Equation 8.6. According to this equation, the optimal incentive for taking action a is the expected value of the state it leads to (i.e., s') minus the expected value of the current state:[2]

$$i^\star(a; s) = \mathbb{E}\left[V^\star(s') \mid s, a\right] - V^\star(s). \tag{8.6}$$

Lieder et al. (2019) tested these optimal incentives in a 10-day-long online experiment. Participants were given a choice between a small immediate reward and the opportunity to earn \$20 by writing five essays by a deadline that was ten days later. Everyone who opted into this task received a to-do list with five writing assignments (see Figure 8.8). Participants were either assigned to one of two experimental conditions, where the optimal incentive for each task was shown next to it on their to-do list, or to one of two control conditions. Participants in the first control condition received no incentives. In the second control condition, each assignment was incentivized by the same number of points. In the first experimental condition, the optimal incentives were shown as dollar values (e.g., \$4.58). In the second experimental condition, they were shown as stars (e.g., 458☆; see Figure 8.8). In both cases, the incentives were just game elements without monetary value. That is, participants could collect points and advance to the next level. But the only money they could earn was the \$20 that everyone received who completed all five assignments on time.

Lieder et al. (2019) found that about 40% of the participants in the control conditions failed to achieve their goal (to earn \$20). Most of them procrastinated so much that they never got started. By contrast, 80% of the participants in the experimental conditions with optimal incentives achieved their goal.[3] This suggests that optimal incentives can help people overcome procrastination in a setting like this. This improvement cannot be explained by the presence of incentives per se because participants in the condition where each task was assigned the same incentive procrastinated just as much as participants in the control condition without incentives. Moreover, in a follow-up experiment, the optimal incentives computed according to Equation 8.6 were significantly more beneficial than an incentive structure designed according to a simple heuristic (i.e., making them proportional to the length of the task). Subsequent work made the algorithms used to compute the optimal number of points more scalable so that they could, in principle, be applied to the kinds of to-do lists people have in the real world (Consul et al., 2022b).

2. A state's value is the expected sum of rewards the person will collect if they take optimal actions starting from that state.

3. Whether the incentives were conveyed by dollar values or stars had no effect on the results.

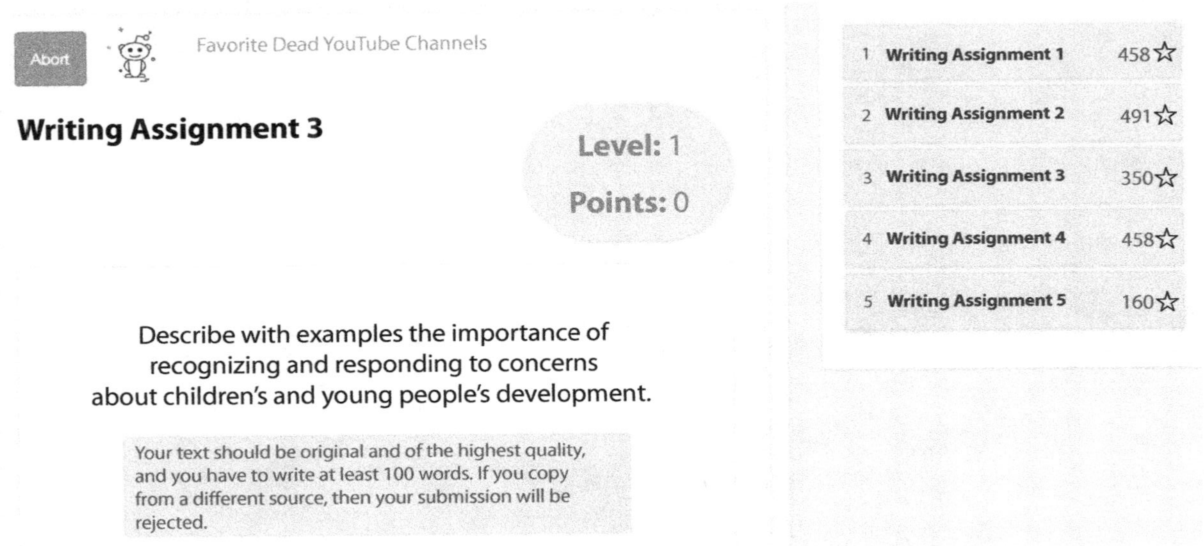

FIGURE 8.8. Screenshot of the to-do list gamification app that Lieder et al. (2019) used to test the optimal incentives derived from a boundedly rational model of decision-making. Here, the incentives were conveyed by the number of stars shown next to each task on the participant's to-do list.

8.6 Resource-rational nudging

When people's innate cognitive capacity is insufficient to solve a problem, we can either increase their cognitive abilities to match the complexity of the problem or simplify the problem to match people's cognitive capacities. Having discussed the former approach (i.e., cognitive augmentation) in the previous section, we now turn our attention to the latter, namely, presenting complex decisions in such a way that people's simple heuristics can produce better choices. This approach is known as **nudging**.

Complex decisions can be presented in numerous ways. Many of them will seem advantageous. But which one is best? If we had infinite time and resources, we could test them all and select the one that improves people's decisions the most. But we often don't have that luxury. Instead, we must select one or a couple of options before their effects are known. So, how can we predict which presentation format is likely to be the most beneficial given what is already known? Applying a theory distilling the findings of many previous studies on the heuristics and biases of human decision-making is one of the most powerful ways to leverage what is already known about people's boundedly rational decision-making. Resource rationality is one such theory. We can, therefore, use it to predict how alternative ways to present a decision might affect people's choices. Moreover, we can use it to find the nudge that would improve people's choices the most if our resource-rational analysis of the decision problem were correct. This is the idea behind **optimal nudging**.

8.6.1 Optimal nudging

Callaway et al. (2023) defined "optimal nudging" as modifying a decision situation such that the choices of a resource-rational decision-maker would achieve the decision architect's goal as well as possible. Formally, adding a nudge n modifies one or more specific situations (s) people encounter in a decision environment (D) from s to $s \oplus n$ such that people using the resource-rational heuristic $h^\star$ will make "better" decisions relative to some goal g. An optimal nudge $n^\star$ does this at least as well as all the other nudges that could have been used instead (see Equation 8.7). The nudge might achieve that by improving the performance of the heuristic the person would have used anyway (e.g., choosing the default option), for instance, by changing the default.[4]

---

4. In the real world, nudging sometimes fails because people distrust the decision architect and seek to protect themselves from manipulation. Our resource-rational analysis of nudging is incomplete because it doesn't take this into account.

Alternatively, it might improve the cognitive operation chosen by the resource-rational cognitive policy from $\pi_{D,C}^*(s)$ to $\pi_{D,C}^*(s \oplus n)$. This can, for instance, be achieved by making some information easier to notice or process. Human behavior is, of course, not perfectly predictable. To account for that, Equation 8.7 takes the expected value over everything a person might do, given that we don't know all the details of the state and the situation (s) that they will be in when they encounter the nudge:

$$n^* = \arg\max_{n \in \mathcal{N}} \mathbb{E}\left[g(\pi_{D,C}^*, C, s \oplus n)\right]. \tag{8.7}$$

In some applications of nudging, the goal is to improve a decision environment to help people achieve their personal goals as successfully and as easily as possible. In this case, nudging amounts to adding nudges ($\mathbf{n}$) to one or more situations in the person's decision environment D to create a better environment $D \oplus \mathbf{n}$. In this case, the objective function of nudging is the resource rationality of the decision-maker (RR). Using the notation introduced in Equation 3.10 from Chapter 3, we can define the optimal way of nudging people to achieve their own goals as

$$\mathbf{n}^* = \arg\max_{\mathbf{n} \in \mathcal{N}^*} \arg\max_{\pi \in \Pi_B} RR(\pi, B, D \oplus \mathbf{n}), \tag{8.8}$$

where $\mathbf{n}$ is a combination of one or more nudges and $\mathcal{N}^*$ is the set of all possible combinations of nudges. Equation 8.8 defines the best way to nudge a resource-rational decision-maker to be more resource-rational. Although this may sound impossible, it is not. The reason is that a nudge can transform a situation in which a resource-rational heuristic fails to choose the action that achieves the decision-maker's goals into a modified situation in which it will select it. For instance, for some environments, the resource-rational heuristic may only consider the first three attributes. If this is so, ensuring that the most important attribute is among the first three attributes will increase the quality of the strategy's decision without increasing its cognitive cost, thereby making the resource-rational strategy even more resource-rational. In some cases, the nudge may render it resource-rational for one to use a more accurate cognitive strategy by reducing its cost. In other cases, adding nudges to most instances of a certain type of decision might change the resource-rational strategy to a simple heuristic that works only because of the nudge. In that case, people might learn to use that simple heuristic instead of their original strategy. For instance, ensuring that the most important attribute is always listed first could make it resource-rational for people to base their decisions on the first attribute only. If the decision-maker cannot tell if a decision situation was modified in this way, then this nudge is only beneficial if the proportion of modified decision situations is large enough to offset the simple heuristic's potential errors in unmodified situations. Our definition of optimal nudging captures these considerations by making the chosen heuristic contingent on its performance in the modified environment (see Equation 8.8).

In other applications of nudging, the choice architect and the decision-maker have different goals. For instance, the choice architect may wish to nudge the person to choose the option that is best for society (e.g., recycling) over the option that is best for the decision-maker's narrow self-interest (e.g., throwing everything into the trash). In that case, the objective function of nudging is the degree to which the choices of a decision-maker using the resource-rational heuristic that maximizes their personal self-interest will achieve the choice architect's goal g (see Equation 8.7 and Equation 8.9).

Equation 8.7 assumes that the decision-maker is resource-rational. If it were known that the decision-maker used a suboptimal cognitive policy π^- instead of the resource-rational policy, then we could obtain better nudges by replacing $\pi_{D,C}^*$ with π^-. This also applies to special cases where the goal is to help decision-makers achieve their own goals. In that case, nudging amounts to modifying the decision situation such that people's heuristics become as resource-rational as possible (see Equation 8.9):

$$\mathbf{n}^\star = \arg\max_{\mathbf{n} \in \mathcal{N}^\star} \mathrm{RR}(\pi^-, \mathcal{C}, D \oplus \mathbf{n}). \tag{8.9}$$

8.6.2   Optimal nudging as a framework for modeling, predicting, and controlling the effects of choice architectures

Callaway et al. (2023) developed a resource-rational framework for modeling, predicting, and controlling the effect of nudges on the decision process. They developed and evaluated this framework in the context of choosing between multiple alternatives based on their attributes. In their experiments, participants chose between bundles of prizes. The prizes differed in value, and each bundle was defined by how many prizes of each type it contained (see Figure 8.10). As illustrated in Figure 8.9, the key idea of their framework is that each decision process comprises a series of cognitive operations that are chosen sequentially. Each of these operations updates the decision-maker's beliefs according to the information it processes. Assuming that people choose these cognitive operations rationally makes it possible to predict what information they will process and what the resulting decision will be. Within this framework, the effects of nudges can be modeled as modifying the decision-maker's initial belief about the alternatives they are choosing between and making some pieces of information easier to process than others. Individual

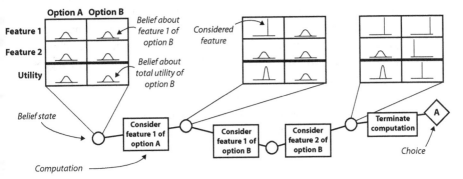

FIGURE 8.9. The resource-rational perspective on nudging models' decision-making as a series of cognitive operations. Each cognitive operation updates the decision-maker's belief based on one piece of information. Each cognitive operation is chosen rationally.

Prizes	Basket 1	Basket 2	Basket 3	Basket 4	Basket 5
A: 3 points	2		3	4	
B: 2 points	7				7
C: 2 points	7	4		2	
D: 21 points	7		8	6	
E: 2 points	9				6

Total click cost: 10 points

You won 2 A prizes, 7 B prizes, 7 C prizes, 7 D prizes, and 9 E prizes, totaling 199 points.
Total earnings (prize values minus click cost): $0.063.

FIGURE 8.10. The Mouselab decision-making paradigm used in the experiments by Callaway, Hardy, and Griffiths (2023)

differences in preferences between different attributes are modeled in terms of the weights different decision-makers assign to each feature.

Callaway et al. (2023) illustrated how their framework can be applied to model the effects of three types of nudges: defaults, suggestions, and highlighting. To model the effect of defaults, the framework assumes that 1) the decision-maker will receive the default option if they don't make a decision, and 2) the decision-maker will update their initial belief about the alternatives according to the assumption that the default option is best for the average consumer, but not necessarily best for their own idiosyncratic preferences. To model the effects of highlighting, the framework assumes that highlighted information is easier to process, thereby reducing the cost of the corresponding cognitive operations. Finally, to model the effect of suggestions, the framework assumes that they have two effects: 1) they add a new alternative to the choice set, and 2) they inform the decision-maker about the best feature of the suggested alternative. With this framework, no prior empirical data on the effects of nudges are required to predict their effects. Moreover, using this framework, optimal nudges can be computed completely automatically.

Across a series of five experiments, the resource-rational framework accurately predicted the effects of each of the three types of nudges. In each case, the resource-rational framework correctly predicted that 1) the nudge improved the quality of participants' decisions, 2) the nudge reduced the amount of information they had to process to reach their decision, 3) the benefit of the nudge increased with the complexity of the decision, and 4) the nudge was most beneficial for people with typical preferences.

Callaway et al. (2023) used their method to automatically compute information-highlighting nudges. They evaluated these optimal nudges against a heuristic approach to nudging (i.e., highlighting the most extreme values) and randomly chosen nudges (i.e., selecting the highlighted values at random). Two experiments confirmed that the automatically computed information-highlighting nudges were more effective than nudges constructed based on the heuristic and highlighting information at random. Concretely, participants supported by the optimal nudges made better choices, which earned them more money, while they exerted less effort.

These findings suggest that the resource-rational framework can be used to design better nudges and understand their effects on decision-making. Whether, when, and for what purpose nudging should be used is a separate question.

8.6.3 Two real-world applications of optimal nudging

Some of the decisions that are most difficult for people involve choices between actions that immediately yield good outcomes and long sequences of unpleasant actions that eventually yield very good outcomes many steps down the road. In such decisions, planning one step or just a few steps ahead favors the actions that yield immediate rewards. Planning many steps ahead quickly becomes computationally intractable because the number of possible action sequences increases exponentially fast with the planning horizon. As a result, people often fail to choose the actions that would be best for them in the long run (O'Donoghue and Rabin, 2015).

One domain in which people often fail to choose what's best for them is their education. Learning valuable skills often requires engaging with difficult content and challenging exercises. In the short run, it would be more enjoyable to play video games or exploit a skill one has already mastered. Digital learning environments often use gamification to keep students engaged. For instance, *Dawn of Civilization* was a game-based digital learning environment that helped youth from developing countries in Asia and Africa learn English. By playing the learning games, the users earned an in-game currency they could invest into building a city. The learning games differed in difficulty, ranging from simple games for practicing skills the player had already mastered to difficult games on skills that the player had yet to learn. The problem was that many players chose to play simple games that allowed them to rapidly earn coins without difficulty—and without any learning. To help students make better decisions about how to spend their time, we applied optimal nudging to the choice of activities in digital learning environments (Pauly et al., 2022; Xu et al., 2019; Pauly et al., 2025). To nudge students to choose the most valuable learning activities in *Dawn of Civilization*, we made them more appealing by adding **brain points**[5] to the screen in which learners select which game to play (Pauly et al., 2025). Brain points are shown on top of the icons of the most pedagogically valuable games. The more the student can learn from the game, the more brain points the student receives. This reframes the choice between games that are hard and frustrating versus games that are fun and easy into a decision between earning many brain points versus earning fewer brain points or no brain points at all. As illustrated in Figure 8.11, the nudge makes it very salient to each student which game they can learn from the most given what they already know. This directs their attention to the best option(s) and makes them more appealing. To determine which games to incentivize with how many brain points, we compute how much each possible sequence of game choices will increase the student's skills and knowledge. Each game is then assigned a number of points proportional to how much it is expected to advance the student's knowledge based on a simple model of learning and the student's current skill levels according to their performance in previous games. We assessed the effectiveness of our nudge in a three-month-long field experiment with 4,410 users of *Dawn of Civilization* from Indonesia, India, Malaysia, Thailand, and a few other countries. Users were randomly assigned to either

---

5. This brain points intervention is both a nudge and an incentive. These two approaches are not mutually exclusive.

FIGURE 8.11. Optimal nudging applied to education. The image is a screen-shot from the digital learning environment *Dawn of Civilization*. The user has to select an educational game. The brain points values (shown in circles on the top right of some games) nudge the user to choose the game that they can learn from the most given what they already know (i.e., "Flying Robot").

see the original version of *Dawn of Civilization* (control group) or the modi-fied version that included our optimal nudge (experimental group). Over the course of three months, users who received our optimal nudge made signifi-cantly more progress in learning English than users of the original version of *Dawn of Civilization* (Pauly et al., 2025).

Health behavior change is another domain in which people struggle to take the actions that are best for them in the long run. For instance, overweight peo-ple could benefit from the habit of drinking a large glass of water before every meal. But to build this healthy habit, they would have to choose this unsavory action over more savory alternatives not just once but for twenty days in a row or longer. Only then, twenty days and many thousands of decisions later, will their initial decision begin to pay off. Planning twenty days ahead is hard, and most people don't do it and consequently fail to choose the healthier option consistently enough to make it a habit. Yet, in retrospect, they would likely agree that it would have been better. To make it easier for people to choose this better option, Lieder et al. (2024b) applied optimal nudging to the recur-ring decision of whether to drink a glass of water at a specific time of the day (e.g., right before lunch). The nudge was delivered by a chatbot that suggested the desired behavior at the chosen time and reframed the decision as a game that posed a much simpler decision: do you want to gain health points or do you want to lose them? When the user chose the water, they gained a number of health points proportional to the health benefits gained by strengthening the habit. When they decided against it, they lost a number of health points proportional to the health benefits they had lost by weakening the habit. The calculation of these point values was informed by a model of habit formation and the habit's current strength according to when the user did versus didn't follow through with the planned behavior (i.e., drinking water) in the past. Lieder et al. (2024b) evaluated this nudge in a 40-day-long field experiment with two active control groups. The results showed that users receiving the optimal nudge chose the healthy behavior on significantly more days than users who received only reminders and users who received reminders and verbal feedback without health points. This suggests that reframing the decision in terms of health points that change people's initial belief about which action is better was effective at nudging them to invest in their health.

8.6.4   Related work, potential future directions, and the ethics of nudging

Nudging is a widely used and increasingly popular approach to improving people's decisions (Thaler and Sunstein, 2021). Nudging has been applied to improve people's personal decisions about their health, wealth, and well-being. Successful examples of such applications include encouraging people to choose healthy foods (Bucher et al., 2016), enrolling them in a retirement savings plan by default (Choi et al., 2004), and reminding them to get vaccinated (Milkman et al., 2022, 2021, 2024). Nudging has also been applied to improve decisions that are important for the health, wealth, and well-being of society. Some successful examples of these applications include making renewable energy the default choice for consumers (Ebeling and Lotz, 2015; Kaiser et al., 2020) and giving them feedback on how their energy use compares to that of their neighbors (Allcott and Kessler, 2019). Many of these applications, including nudges to save energy and get vaccinated, benefit both the individual and society at the same time.

Three of the most popular types of nudges are defaults, suggestions, and information highlighting. When experts, practitioners, and policymakers try to predict which nudges will be effective based on intuition or qualitative principles, those predictions can be surprisingly inaccurate (Sunstein, 2017). For instance, both policymakers and behavioral scientists predicted that offering free rides to vaccination sites would boost vaccination rates, but this nudge turned out to be completely ineffective (Milkman et al., 2024). We hope that the resource-rational framework of optimal nudging introduced in this section will enable more accurate predictions of which nudges will be effective. The findings from a series of online experiments by Callaway et al. (2023) suggest that this is a promising approach. However, how well this approach works in the real world is still unknown. This makes evaluating how accurately resource-rational models can predict the effectiveness of nudges in the real world an important direction for future research. Relatedly, the field experiments presented above suggest that deriving nudges from resource-rational models could be a promising approach to improving decision-making in the real world. However, more research is needed to more accurately evaluate the promise and boundary conditions of this approach.

As nudging has become increasingly popular in public policy, its ethicality has come under scrutiny (Hertwig and Grüne-Yanoff, 2017; Schmidt and Engelen, 2020; Bovens, 2009; Chater and Loewenstein, 2023). The most prevalent ethical concern about nudging is that nudges undermine people's autonomy by manipulating their choices without their consent (e.g., Goodwin, 2012; Wilkinson, 2013; Jung and Mellers, 2016). Such concerns can sometimes be mitigated by being fully transparent about the nudge, its intended purpose, and the costs and benefits of relying on it (Sunstein, 2015). Another important insight is that the ethicality of nudging—like the ethicality of laws prohibiting certain behaviors—depends on the purpose of the nudge (e.g., preventing murder versus luring people into a scam) and the details of its implementation (e.g., transparency and consent) (Sunstein, 2015). It seems reasonable that certain applications of nudging should only be allowed when an independent ethics commission can verify that the risks for the individuals being nudged are much smaller than the expected benefits that the nudge will have for them and society at large (Sunstein, 2015). Section 8.8.2 outlines a theoretical framework that can be used to design nudges that have that property.

8.7   Rationality enhancement as a combinatorial optimization problem

The approaches to improving decision-making reviewed above are not mutually exclusive. Boosting, nudging people to be more resource-rational, and cognitive augmentation are complementary strategies for enhancing human rationality. Rationality enhancement can combine all of these approaches. In some cases, combinations of multiple approaches might be able to improve certain decisions significantly more than any single approach. For instance, in some cases, giving people new computational tools (cognitive augmentation) can significantly increase the range of cognitive strategies people can execute, thereby creating new opportunities for boosting. To realize this opportunity, optimal boosting can be applied to a model of the augmented mind. We might be able to do even better if the computational tools are designed with boosting in mind. That way, we might discover thinking tools that create genuine synergies between human and machine intelligence. In principle, the optimal combination of cognitive augmentation and boosting might be completely different from what we might get by conducting boosting and cognitive augmentation independently and then combining the results. This suggests we should approach rationality enhancement as one big combinatorial optimization problem, rather than as a collection of complementary optimization problems.

According to this view, optimal rationality enhancement consists in providing people with the best possible combination of boosts, nudges, and cognitive augmentation (see Equation 8.10):

$$(i^\star, n^\star, c^\star, h^\star) = \underset{(i,n,c,h) \in \mathcal{I} \times \mathcal{N} \times \mathcal{C}_{\text{ext}} \times \Pi_{\mathcal{C} \oplus c}}{\arg\max} \mathrm{RR}(\pi, \mathcal{C} \oplus c, D \oplus n \oplus i). \quad (8.10)$$

Finding ways to solve this combinatorial optimization problem, characterizing its solutions, and testing them in practice are interesting directions for future research.

8.8 Conclusion and future directions

8.8.1 Summary and implications

As laid out above, the theory of resource rationality can be used to ground various approaches to improving people's decisions in computational models of human cognition. This makes it possible to understand and predict the effects of alternative interventions on human decision-making. Moreover, the theory of resource rationality makes it possible to derive interventions for improving people's decisions from principled, evidence-based assumptions about people's cognitive capacities and their limitations. As illustrated above, it is, in principle, possible to compute which nudge, boost, or incentive would be most beneficial according to a model of the decision environment, a model of the decision-maker(s), and a utility function that quantifies the desirability of alternative possible outcomes.

The approaches described above are still very new. Therefore, as one might expect, their applications have been limited to online experiments and proof-of-concept studies. The results of most of these studies were very encouraging. Extending theories, computational methods, and theory-based interventions from the laboratory to the real world is always challenging.

Applying rationality enhancement to the real world certainly won't be easy. Obtaining sufficiently accurate models of real-world decisions and decision environments will definitely be challenging. No matter how rigorously we measure and model the structure of real-world decisions, there is always a risk that something important will be overlooked or misestimated. An intervention that is optimal, according to an inaccurate model of the world, could, in principle, do more harm than good. Real-world applications of rationality enhancement should, therefore, be developed very carefully and tested rigorously before they are rolled out on a large scale. However, these risks are in no way unique to the resource-rational approach. Any approach to designing decision environments, incentives, and interventions—including the theories that our current systems are based on—is subject to those risks. For our approach, these risks can be partially mitigated by robust methods that optimize boosts, nudges, and incentives with respect to a model that takes those risks and uncertainties into account (Mehta et al., 2022). Developing sufficiently accurate models and extending and scaling robust methods to real-world applications of rationality enhancement will require a large collaborative effort over an extended period of time and significant investments. However, we believe that the potential benefits for society would make these efforts an extremely worthwhile investment in the future of humanity. We excitedly look forward to future work on extending the principles of rationality enhancement toward real-world applications. In the remainder of this section, we point out some exciting long-term directions for improving real-world decisions that could be pursued within the resource-rational framework in the coming decades and centuries.

As we ponder what remains to be done, one gap stands out. Our research so far lays the foundations for helping individuals pursue their interests more rationally. However, individuals' interests are not always aligned with the greater good. Enhancing people's ability to achieve goals that are misaligned with the greater good could allow them to cause great harm. Most of the approaches presented above do not take this into account. Therefore, future work should extend rationality enhancement from applications that benefit individuals to applications that benefit society.

8.8.2 Beyond rational self-interest

Incentives, nudges, and boosts are often designed to benefit the greater good. To apply resource-rational analysis in the service of this noble goal, we have to extend the resource-rational approaches presented above from helping the decision-maker to helping society. While the applications presented above brought people closer to the ideal of rational self-interest, future work should apply these approaches to bring people closer to the ideal of rational altruism. We won't claim that people are ever completely altruistic, let alone completely rational and completely altruistic. But we will argue that deriving interventions from the ideal of rational altruism might be a promising way to increase the proportion of people's decisions that benefit the greater good and the degree to which they benefit it (see Table 8.1). We believe the pursuit of this goal will be more successful when it aims at a realistic target—resource-rational altruism—and builds on realistic models of the human mind derived through resource-rational analysis.

The resource-rational approach described in this chapter can also be applied to foster decisions that benefit the greater good. This can be achieved by computing the intervention that maximizes expected utility minus expected cognitive cost, as detailed above. The only difference is that the utility function measures the utility for everyone rather than only for the decision-maker. Two use cases are worth distinguishing between: 1) helping altruists pursue their altruistic goals more effectively, and 2) guiding less altruistic individuals to socially beneficial decisions. The latter includes guiding self-interested individuals to decisions that will benefit the greater good and away from decisions that are net-harmful for their group or society. It also involves guiding individuals with mixed motives, some more selfish and some more prosocial, to act more on their prosocial motives and less on their selfish motives.[6]

The first use case is straightforward and relevant. It is relevant because almost everyone has prosocial motives, but almost nobody pursues them rationally (Caviola et al., 2021). Many people regularly make decisions about how to pursue their prosocial motives. This includes decisions about whom to help and how to help them. For instance, many people regularly decide which charity or charities to donate to and how much money to donate to each of them. Unfortunately, the heuristics they use are often highly suboptimal. For instance, donors often rely entirely on intuition and the charity's emotional appeal, even though this heuristic works very poorly in the context of charitable giving (Caviola et al., 2021). Teaching donors better decision strategies, that is boosting, could help mitigate this problem to some extent (Caviola et al., 2020). For instance, donors often pay too little attention to crucial information about how many people will benefit and how much they will benefit. Moreover, even when donors attend to information about the charity's effectiveness, they give too much weight to the charity's administrative expenses and too little weight to its cost-effectiveness. These and other shortcomings for deciding between charities can be addressed by teaching them

---

6. The same logic also applies to motives that are antisocial.

Table 8.1. Some opportunities for future work on resource-rational approaches to improving decisions for the benefit of the greater good.

Method	Example Application
Optimal boosting	Deriving resource-rational heuristics for prosocial decision-making.
Optimal boosting	Deriving optimal moral rules and heuristics for resource-rational altruism in social dilemmas.
Optimal boosting	Leveraging resource-rational models of decision-making to derive optimal decision rules and decision procedures for societal decision-making (e.g., voting systems).
Systematic reflection	Designing Socratic questions that help people discover resource-rational decision-strategies for fostering the greater good.
Optimal nudging	Using resource-rational models to compute nudges that maximize the positive social impact of people's charitable donations (e.g., nudging donors to give to effective charities).
Optimal nudging	Using resource-rational models to compute nudges that maximize the probability that people will constructively contribute to group decision-making (e.g., share crucial information and base arguments on reason and evidence) and societal decision-making (e.g., vote).
Optimal nudging & optimal incentives	Using resource-rational models to compute nudges and incentives that maximize the probability of successful cooperation in social dilemmas.
Optimal incentives for resource-rational agents	Leveraging resource-rational models of decision-making to design incentive systems that align the decisions of governments and other elected or appointed officials with the interests of the people they represent.
Optimal incentives for resource-rational agents	Designing incentive systems that make effective charities the most appealing option for donors with impure motives, limited information, and cognitive biases.
Cognitive augmentation	Leveraging resource-rational models of cognition to derive decision support systems that enable groups to make more rational decisions on behalf of society.
Resource-rational framing	Deriving resource-rational representations of social dilemmas that enable people to understand their decisions' most important consequences.

decision strategies that focus on the charities' cost-effectiveness. Changing the way in which donation decisions are framed and how the charities are presented can also have a powerful effect on people (Chan et al., 2024; Saeri et al., 2023; Caviola and Greene, 2023). That is, charitable giving can be improved through nudging. Finally, there is a substantial literature on how people's donation decisions are influenced by motives other than altruism, such as the desire to feel good, the desire for social recognition, and other personal objectives (Bekkers and Wiepking, 2011). These incentives are often misaligned with the charities' effectiveness, suggesting that designing better incentives could make people's donations much more effective.

Compared to the extensive work on interventions to foster charitable giving (Chan et al., 2024; Saeri et al., 2023), the amount of research on promoting rational decisions about which charity to give to is very limited (Caviola et al., 2021). So far, efforts to boost, nudge, and incentivize effective giving have been guided by qualitative theories about heuristics and biases. The effectiveness of these efforts has been mixed, and there is likely significant room for improvement. None of the existing interventions were derived using the methods of optimal nudging, optimal boosting, or optimal incentive design presented in this chapter. Applying these methods might allow us to create boosts, nudges, and incentives that promote effective giving much more effectively. Table 8.1 summarizes several concrete opportunities for future work on resource-rational nudging and boosting to help people overcome obstacles to effective altruism.

Charitable giving is not the only way in which people can pursue their prosocial motives (Lieder et al., 2022b). People can also do good by working directly on important causes, such as sustainability, ensuring that advanced AI will be safe and socially beneficial, and preventing future pandemics. Unfortunately, the lack of information, the misaligned incentives, and the cognitive biases that prevent people from choosing effective charities, often also prevent them from choosing an impactful project with a higher positive social impact. Applying the principles outlined in this chapter to help people rationally consider their expected social impact in career decisions and project selection is an important direction for future work. Fortunately, recent work has already produced some actionable recommendations for psychologists and other decision scientists (Gainsburg et al., 2023; Kassirer et al., 2023).

The second use case—guiding less altruistic individuals to altruistic decisions—is only slightly more challenging. The main difference is that the utility function used to predict the intervention's effect on the person's decision is different from the objective function that the intervention is designed to optimize. The objective function weighs the utility of all stakeholders equally, whereas the utility function used to model the decision-maker's preferences gives more weight to their own welfare and the welfare of people close to them. We have already incorporated this distinction into our definitions of optimal incentives (Equation 8.3) and optimal nudges (Equation 8.7). Extending optimal boosting to prosocial decisions is a promising direction for future research. Teaching virtuous heuristics and moral rules derived from a resource-rational analysis of prosocial decision-making could become promising new approach to character education and moral education (see Table 8.1).

Designing incentives and nudges to guide self-interested individuals toward prosocial decisions has a rich history in economics (Wu et al., 2022; Bénabou and Tirole, 2006; Capraro et al., 2019; Bao and Ho, 2015; Carlsson and Johansson-Stenman, 2019; Ghesla et al., 2019). However, as far as we know, none of these approaches have leveraged resource-rational models of cognition. Instead, these approaches were either based on the homo economicus model, qualitative models of heuristics and biases, or coarse mathematical abstractions. We propose that these approaches can be significantly improved by instead deriving nudges and incentives from realistic process models of human cognition that accurately capture people's cognitive mechanisms and their limitations. This will be a substantial undertaking that will require many iterations of basic, translational, and applied research. In addition to the scientific and technical challenges of this work, there are also a number of practical, legal, and political issues that will need to be addressed to ensure that the resulting applications are ethical.

8.8.3   Beyond individual decision-making: Improving the decisions of groups, organizations, and society

Many important decisions are made by groups, organizations, governments, or society as a whole. Under some circumstances, groups are more rational than the individuals who comprise them (Kugler et al., 2012; Curşeu et al., 2013). However, under many other circumstances, group decision-making is afflicted by many biases that can cause groups to make worse decisions than individuals (Kerr and Tindale, 2004; Jones and Roelofsma, 2000). For instance, once one person has shared an initial idea, everyone else in the group becomes more likely to share information that supports that idea than information that contradicts it. This **groupthink** causes many groups to become highly overconfident in the initial idea and fail to consider information that speaks against it and other alternatives. As a result, group decisions are often more extreme than what any individual member of the group would have chosen.

These failure modes of group decision-making are not inevitable. The process of group decision-making can be structured to follow decision procedures that can enable the group to be more rational than its members (Sunstein and Hastie, 2015). For instance, the problem of groupthink can be overcome by using a structured, three-step process (Girotra et al., 2010). In Step 1, each individual generates a ranked list of proposals independently of everyone else. In Step 2, the group discusses all ideas generated in Step 1 and generates additional ideas. In Step 3, the group reaches a consensus on a ranked list of five best ideas out of all ideas generated in Steps 1 and 2. Such structured decision processes can be facilitated using technology (Miranda, 1994).

So far, strategies for enhancing the rationality of group decision-making have been designed based on qualitative theories and improved through trial and error. We believe that this decision process could be improved by leveraging agent-based models of group decision-making that model the behavior of each individual according to the best available models of the underlying cognitive processes, including resource-rational models of individual decision-making. These models should include social motives, such as the desire for social approval, and relevant cognitive biases, such as the anchoring bias (Lieder et al., 2018b). In this way, the resource-rational framework could be used to derive structured procedures for group decision-making and societal decision-making. Leaders and managers could be provided with step-by-step guides for facilitating group decision-making. Alternatively, a decision-support system could guide groups to follow resource-rational procedures. In addition, workshops and tutorials could be conducted to teach people how to use these tools and procedures. Moreover, nudges could be designed to increase team members' compliance with these more effective procedures. Finally, the resource-rational framework could also be used to compute optimal incentives for encouraging individuals to contribute to group decisions in the most constructive way possible. Such incentives could help correct the unhelpful incentives causing groupthink or compensate for them. Moreover, optimal incentives could counteract the lack of sincere engagement that often results from the diffusion of responsibility associated with group decision-making.

The most challenging decisions groups and societies face are **social dilemmas** (Dawes, 1980). Social dilemmas have three defining properties. First, the interests of different stakeholders are in conflict with each other. Second, if everyone acts on their rational self-interest, everyone will be worse off. Third, everyone is better off if everyone cooperates with the group. There is a considerable amount of work on how social dilemmas can be solved (Kollock, 1998). Many of these solutions derive incentives for cooperation, including rewards for prosocial behavior and punishments for antisocial behavior, from the homo economicus model (Börgers, 2015). Other interventions are grounded in psychological research on human decision-making in social dilemmas (van Lange et al., 2013). Recent work has used methods from reinforcement learning to learn solutions to social dilemmas (e.g., Koster et al., 2022). However, none of the existing approaches was explicitly derived from realistic process models of human decision-making in social dilemmas. Applying the resource-rational framework to solve social dilemmas is a promising direction for future work (see Table 8.1).

Some of the most important societal decisions are, directly or indirectly, made through elections and referendums. Research in social choice theory strives to improve such decisions by designing optimal voting systems and related mechanisms (Sen, 1986). Most work on social choice theory assumes that the voters are fully rational. In recent decades, behavioral choice theorists have realized that their field's theoretical recommendations depend on rationality assumptions that are often empirically false (Regenwetter et al., 2009). We believe that resource-rational models of decision-making can help the next generation of social choice theorists rebuild their field on a more solid foundation comprising empirically supported models of human decision-making. Facilitating these efforts by developing and testing resource-rational models of voting is an important direction for future research on decision-making. Subsequent research should develop resource-rational models of voting that can accurately predict the social welfare consequences of alternative voting systems and apply them to derive optimal voting systems (see Table 8.1). Although this work is clearly very challenging, it could become important for the future of democracy.

One unique challenge of improving the decisions of groups, organizations, and societies is that they result from the interactions of many different actors, some of whom are more powerful than others. Therefore, to find the most effective solutions to societal problems, we should not only consider the full range of possible interventions but also the full range of decision-makers whom our interventions could target. Broadening our perspective in this way can help us overcome the blind spots of more narrow approaches to improving decisions, such as nudging. For instance, some people are concerned that politicians' overly optimistic views about the power of nudging prevent them from implementing more effective solutions to systemic problems, such as regulation and taxation (Chater and Loewenstein, 2023). Chater and Loewenstein (2023) argue that the current emphasis on nudging falsely shifts the responsibility for systemic problems, such as climate change and the opioid epidemic, to individual consumers even though the root cause of those problems is the lack of regulations and incentives that align corporate interests with the greater good. Applying rationality enhancement to societal decisions about issues such as climate change could provide us with a principled approach to identifying the most effective ways to improve how humanity will make some of its most impactful decisions.

In conclusion, extending resource-rational boosting, nudging, and incentive design to group decision-making is an important direction for future research on the crucial problems of improving societal and organizational decision-making. Applying rationality enhancement to organizational and societal decisions will require a precise understanding of those complex social situations and how people think about them. Understanding those situations well enough that rationality enhancement can be applied to them will likely take many decades, if not centuries, of research. Nevertheless, there are economic games and social dilemma paradigms that are likely already within reach for resource-rational analysis or will be within reach in just a few years or decades.

8.8.4   Beyond cognitive strategies: Improving mental representations

The interventions discussed so far target the process of decision-making. Another prerequisite for making good decisions is to use good representations. Indeed, previous research found that improving people's representations can be very effective at helping people to think more rationally (Hoffrage and Gigerenzer, 1998; Hoffrage et al., 2002), make better decisions (Hershfield et al., 2011), and solve complex problems (Patrick and Ahmed, 2014; Nutt, 1993).

The recent work summarized in the previous chapter has extended the theory of resource rationality to mental representations, including people's mental representations of the decision problems they are trying to solve. People's representations often leave out some aspects of the situation and emphasize others. On the one hand, this enables them to reach decisions more efficiently with less cognitive effort (Ho et al., 2022). On the other hand, it can sometimes cause them to overlook some important factors, such as side effects (Prystawski et al., 2022) as well as indirect consequences and their actions' effects on distant others (Burga et al., 2023). Sometimes, this causes people to make big mistakes they could have avoided if they had represented the decision differently (Prystawski et al., 2022).

In such cases, their initial representation might not be resource-rational. When that's the case, people's decision-making can be improved by guiding them toward a representation that is more resource-rational, for instance, by directing their attention to crucial facts they overlooked or by asking them to look at an interpersonal conflict from the other person's perspective. Another related approach to improving decision-making is framing. How a decision is framed determines how people will represent it, which in turn determines how likely they are to arrive at different decisions. Future work could apply our framework to find out how a decision has to be framed so that the choices of resource-rational decision-makers will be most beneficial to society (i.e., **resource-rational framing**). This approach could improve people's decisions in many important contexts, including those of making intertemporal choices, navigating social dilemmas, risk-taking, making donations, planning, and problem-solving. Improving people's representations might also be crucial for promoting social mindfulness and prosocial, sustainable, and considerate decisions.

Helping people construct resource-rational mental representations could be a promising approach to improving the extent to which people's decisions are rational and prosocial. The resource-rational framework could provide a principled approach that could enable substantial improvements over existing approaches. If we knew what the resource-rational representations of complex social dilemmas are, it would be much easier to help people think about them more rationally.

8.8.5 Concluding remarks

The mathematical formalism of resource rationality is a helpful framework for the main approaches to improving human decision-making. The resource-rational framework has a number of advantages over current approaches to improving people's decisions. Its advantage over the approaches of traditional economics and operations research is that it is informed by realistic models of the cognitive processes that generate people's decisions. Its advantage over current and previous approaches rooted in psychology and behavioral economics is its use of precise computational models and mathematical optimization techniques. Prior work has demonstrated the approach's feasibility in laboratory paradigms. In the long run, improved versions of the methods outlined in this section might make it possible to extend rationality enhancement to important problems people face in the real world. Immediately applying the findings surveyed in this chapter to the real world would be premature. The choice of such applications should be informed by a careful assessment of potential risks and ethical issues. Finally, incentives, boosts, nudges, and decision support systems derived from resource-rational models of cognition can be used not only to benefit individuals—they can also be used to benefit society at large. The ideas laid out in this section provide a roadmap for how we might get there (see Table 8.1). It is going to be a long and arduous journey along roads that have yet to be built. But it might lead us to a more reasonable future where the decisions of individuals, organizations, governments, and society are better for everyone.

9

Conclusion

In the preceding chapters, we have presented a different way of thinking about rational action, and a method for using this perspective to make sense of human behavior. At the heart of this approach—resource-rational analysis—is the idea that thought itself is subject to rational analysis (Chapter 3). To think well, people have to make a series of good (metacognitive) choices of what to think about next and how to think about it. As illustrated in the previous chapters, using this approach has several advantages. It often allows us to explain and predict when and how people deviate from classical rationality (Chapter 4). It answers the question of how people should choose which strategy to follow to make a decision or solve a problem (Chapter 5). It can also be used to discover cognitive strategies (Chapter 6) as well as a new theoretical framework for answering questions about the representations and cognitive architectures that support planning and decision-making (Chapter 7). Moreover, thinking about human cognition in these terms gives us new tools for helping humans make better decisions—and make better use of their cognitive resources more generally (Chapter 8).

The findings surveyed in this book suggest that resource-rational analysis is a useful method that can enable more accurate predictions and more effective interventions than traditional theories of rational action. This does not require that people are fully resource-rational, and we don't claim that they are (Lieder and Griffiths, 2020a). So far, most work on resource rationality has focussed on developing the methodology of resource-rational analysis and applying it to create useful models of cognition. Testing whether and, if so, how people's cognitive strategies deviate from their resource-rational counterparts is a promising direction for future research. To what extent people are resource-rational remains to be determined. Some initial findings suggest that people's cognitive strategies aren't always fully resource-rational (Krueger et al., 2024; Callaway et al., 2022b,a; Consul et al., 2022a). Krueger et al. (2024) conducted the most thorough investigation of this question to date. They found that people's performance across a range of multi-alternative risky choice problems was 38% lower than the performance of the resource-rational model. The primary reason for this performance gap was that people didn't always choose the most adaptive strategy. This line of research is still in its infancy. Many more studies are needed before we can confidently estimate how resource-rational people are, how human cognition deviates from resource rationality, and how these deviations depend on the task and the structure of the environment. In the long run, this line of research might reveal systematic deviations from resource rationality that can be explained by an even more accurate theory. However, for now, resource-rational analysis is the most useful refinement of previous theories of rational action available to date.

In the remainder of this final chapter, we consider some of the broader implications of this approach, as well as some of the opportunities it creates for future research. Resource-rational analysis is informed by computer science and psychology, but questions about effective decision-making arise in a wider range of disciplines, including economics and neuroscience, and their answers impact topics as far afield as law and policy. While we have begun to explore some of the implications of this approach, realizing its full potential will require integrating resource rationality into the foundational theories of these other disciplines.

9.1   Implications for other disciplines

9.1.1   Resource-rational analysis and the foundations of decision science

Decision science is a multidisciplinary field that analyzes how decisions should be made, how decisions are in fact made, and how decision-making can be improved. Resource-rational analysis potentially provides a unifying theoretical framework for research on all of these questions. Across the chapters of this book, we have shown how resource-rational analysis can help researchers understand, predict, and improve many different types of decisions and related cognitive functions. Resource-rational analysis allows researchers to make principled assumptions about the many unknowns that shape people's decisions. The resulting mathematical models make precise predictions that can be tested empirically.

The field of decision science was built on the foundation of rational theories of optimal decision-making, such as expected utility theory, and empirical behavioral research on actual decision-making. Resource-rational analysis can be seen as a synthesis of the theoretical and empirical foundations of decision science. Like traditional normative models, resource-rational models are based on a rigorous and broad theoretical framework. However, unlike these traditional models, resource-rational models are also directly informed by empirical results. Resource-rational analysis, therefore, has the potential to combine the strengths of two approaches that have traditionally been in tension with one another. In addition to unifying these two key elements of decision science, resource-rational analysis provides connections to research on the architecture, processes, and representations of the human mind. This makes resource-rational analysis a potentially valuable building block in the foundation of decision science.

9.1.2 Implications for economics

Explaining, predicting, and improving decisions are central tasks of contemporary economics. Economists' success at these tasks depends on the accuracy of their models of people's decisions. These models have become increasingly accurate over time. Expected value theory was refined into expected utility theory (von Neumann and Morgenstern, 1944), which was later refined into prospect theory (Kahneman and Tversky, 1979). The descriptive accuracy of psychological science has started to inform economics with the rise of behavioral economics (Thaler, 2015). More recently, insights about cognitive mechanisms and representations have begun to enter economics with the rise of cognitive economics (Enke, 2024; Walliser, 2007). Resource rationality is a promising framework for integrating these new perspectives into economic theory.

The advance from notions of rational self-interest in (neo)classical economics to resource rationality in cognitive economics has many crucial implications for many areas of economic theory, ranging from microeconomics to macroeconomics, finance, public economics, and industrial organization (Gabaix, 2019, 2023). These conceptual changes have implications for monetary and fiscal policy (Gabaix, 2020), taxation (Farhi and Gabaix, 2020), and economic growth (Gabaix, 2023). The project of rebuilding economics on a solid cognitive foundation is promising and very important (Walliser, 2007; Chater, 2015; Enke, 2024).

Despite this progress towards theories of human decision-making that are more realistic than the homo economicus model traditionally assumed in economics, existing models still make relatively simplistic assumptions about the underlying neurocognitive mechanisms and constraints. Therefore, this promising project is still far from complete (Rostain, 2000). The resource-rational models presented in this book are just some of the first steps of a long march toward accurate models of how people make complex decisions. We hope that economists and other cognitive scientists pursuing this goal will take advantage of and benefit from resource-rational models of cognition and the methods of resource-rational analysis and automatic strategy discovery surveyed in this book.

9.1.3   Implications for law and policy

The purpose of policy is to guide the decisions people make within an organization, government, or society. Laws also regulate the behavior of individuals and organizations for the benefit of all. The success of policies and laws crucially depends on how people respond to them. Therefore, the better we can predict people's responses to potential laws and policies, the easier it will be to design them well.

The design of laws and policies is already informed by theories of human behavior. Some of these theories are based on traditional notions of rationality. Unfortunately, people's reactions to new laws and policies often defy the principles of rationality (Sunstein, 2000). This observation and other empirical demonstrations of human irrationality have galvanized the development of the field of **behavioral law** (Sunstein, 2000). Behavioral law strives to rebuild the foundations of lawmaking using empirical findings from behavioral economics. Resource-rational analysis can be leveraged for this project in two ways. First, rational theories can be upgraded to resource-rational theories that make realistic assumptions about people's cognitive limitations. Second, resource-rational analysis can be used as a principled method for translating a wide range of potentially policy-relevant empirical findings from psychology and behavioral economics into mathematical models that can be used to design optimal policies and predict what effects potential laws might have on people's behavior.

In addition to rational theories and empirical findings, lawmakers' and policymakers' frameworks are often based on intuitive heuristics (Sunstein, 2005, 2003). Resource-rational analysis can be used to understand when those heuristics are appropriate and when they are not. Moreover, resource-rational analysis can be leveraged to boost policy decisions by providing policymakers and lawmakers with better heuristics and decision-support systems. By making more realistic assumptions about what lawmakers and policymakers are capable of, the approach to improving decision-making that results from resource-rational analysis can provide them with simple tools that they might actually use.

9.1.4   Implications for neuroscience

Our focus in this book has been on understanding human behavior, but the topics we have considered—decision-making and planning—are the subject of equally rich literatures in neuroscience. We see resource-rational analysis as having exciting potential to guide research in neuroscience. Research on the neuroscience of decision-making has long been guided by the homo economicus model (Glimcher and Fehr, 2013). This led neuroscientists to search for brain areas representing the basic components of expected utility theory: probabilities and utilities. However, as we have seen in Chapter 4, the actual mechanisms of human decision-making might look nothing like the homo economicus model. Therefore, some efforts to find the neural representations of the various variables assumed by economic theory were likely misguided.

On average, neuroscientists searching for the neural basis of some assumed representations and processes will be more successful when those assumptions are actually true. Resource-rational analysis can provide them with better-informed guesses about which representations and processes are worth looking for in the brain. We, therefore, predict that the search for the neural basis of decision-making will likely become much more efficient and more successful once neuroscientists start looking for the neural basis of resource-rational representations and strategies. Moreover, by predicting that decision-making should involve a wider range of cognitive processes and representations than those assumed in (neo)classical economics, resource-rational analysis provides a new set of opportunities to identify underlying neural mechanisms.

In particular, by identifying what cognitive processes people should be engaged in when solving specific problems, resource-rational analysis provides a new set of targets that can be used when trying to decode neural signals. Probability and utility are aspects of decision problems that we can correlate against neuroimaging data to find the neural correlates of decision-making. In the same way, the specific computational actions prescribed by a resource-rational analysis can be correlated against neural signals as a source of clues about the cognitive processes that people engage in and how those differ across tasks and problems. By making precise predictions about these processes, resource-rational models give us new things to look for in the brain.

Another avenue by which ideas from neuroscience can be connected directly to resource-rational analysis is via the formulation of strategy discovery as a sequential decision problem. Sequential decision problems have been studied extensively in neuroscience, with Markov decision processes (MDPs) forming the foundation for research on how humans and other animals learn from environmental rewards (Dayan and Daw, 2008). This literature has identified different kinds of algorithms that people use when learning to solve MDPs and areas of the brain that seem to represent the quantities computed by those algorithms (Dolan and Dayan, 2013; O'Doherty et al., 2015; Lee et al., 2012; Mattar and Lengyel, 2022; Miller and Venditto, 2021). We see this literature as one that can be directly applied to understanding the neural basis of metacognitive learning (Lieder et al., 2018d; Lieder and Griffiths, 2017; Krueger et al., 2017; He et al., 2021; Jain et al., 2019), exploring the possibility that the same brain areas encode similar quantities for internal computations and for external actions. In this way, computational models of metacognitive learning provide cognitive neuroscience with a powerful set of new tools for understanding how people learn effective cognitive strategies.

9.1.5 Implications for artificial intelligence

The roots of resource-rational analysis lie in the artificial intelligence literature, in work on bounded optimality and rational metareasoning that was done as researchers began to think about ideal behavior for agents with finite time and hardware-constrained computational resources (Horvitz, 1987; Horvitz et al., 1989; Russell, 1997; Russell and Subramanian, 1995, Chapters 2 and 3). However, recently, the artificial intelligence literature has moved in a different direction, being less concerned with theoretical questions about ideal behavior and more focused on pragmatic questions about how to create agents with specific competencies. Much of this work has been done in a setting where scarce resources are not a salient constraint: modern AI systems are the result of training on massive amounts of data using massive amounts of computation that may consume about 100,000 times as much energy as the human brain. This paradigm is quite different from that which shapes human cognition (Griffiths, 2020).

Despite this, there has been work within AI that has explored problems related to metareasoning, albeit in isolation from previous literature. For example, Graves (2016) created a neural network that was trained to make a decision about how many iterations of recurrent computation to perform, an idea that was expanded upon by Banino et al. (2021). In computer vision, Teerapittayanon et al. (2016) investigated a strategy whereby a classification decision could be made from early layers of a neural network—reducing computational cost—if there was already sufficient confidence in the answer. As we hit the limits of the computational resources that can be allocated to creating AI systems, we can anticipate that metareasoning is likely to return to the spotlight.

In addition, there are other contexts where the kind of metareasoning we have focused on here are relevant to contemporary AI systems. For example, large language models have proven extremely effective at solving certain kinds of problems, but also face challenges in cases where solutions require multiple steps or advanced planning. One way to address this weakness is to build a cognitive architecture around those large language models in a way that makes it possible to use insights from classical AI (Sumers et al., 2023). For example, classical search algorithms can be guided by the responses from a large language model, resulting in a system that has far greater flexibility in the problems that it is applied to than a traditional AI system but higher performance on complex tasks than a large language model alone (Yao et al., 2023; Piriyakulkij et al., 2024). These connections open the door to further applications of methods related to resource-rational analysis, as research on human cognition continues to reveal principles we can implement in machines.

9.2 Future directions

9.2.1 Solving more complex problems

Metacognition is hard. The difficulty of finding the optimal solution to a Markov decision process increases with the size of its state space. In the problems we have considered, considering the metalevel problem can result in an exponential increase in the size of the state space. For example, in the planning task explored in Chapter 6, the planning problem itself involves just 16 states. The metalevel MDP, however, has one state for each of the 4^{16} possible sets of beliefs that the planner could arrive at. As a result, it has more than four billion belief states. This challenge motivated the development of new algorithms for approximately solving meta-MDPs outlined in the appendix.

The challenge of finding approximate solutions for large meta-MDPs is a limiting factor in applying resource-rational analysis to more complex tasks. However, we are optimistic that this can be done, drawing on the significant progress in solving challenging reinforcement learning problems in the AI literature. Methods based on deep neural networks have made it possible to solve problems that would previously have been considered impossible to solve, and it may be possible to adapt these new methods to meta-MDPs. Exploring whether these methods can be used to develop more efficient algorithms for identifying good solutions to metacognitive problems is a critical direction for future research.

9.2.2 Discovering richer strategies

The kinds of cognitive strategies that we have focused on in this book are relatively simple: parameterized strategies where the optimization problem reduces to choosing a number of samples or a number of steps of computation (Chapter 4), choosing from a fixed set of given strategies (Chapter 5), and strategies that are built step-by-step from a fixed set of elementary operations

(Chapter 6). But human metacognition is far richer than this. We can recognize when a problem has shared components with a problem we have solved before, intuit procedures that could be modified to address a new problem, and reuse parts of strategies we have previously developed.

These aspects of human strategy discovery can potentially be captured by drawing on the connections between resource-rational analysis and reinforcement learning. Specifically, since we can formulate strategy discovery in terms of solving a meta-MDP, we can use ideas about how to learn more structured policies for MDPs as a guide for how to capture more sophisticated forms of human metareasoning. In particular, the literature on **hierarchical reinforcement learning**—mentioned in Chapter 7—has explored how adaptive agents might seek high-level actions ("options") that make it easier to pursue subgoals. The same approach can be applied in a metacognitive setting, where options become components that are reused across different strategies because they provide an effective way to solve a recurring sub-problem.

Another relevant idea from the reinforcement learning literature is that of viewing **policies as programs** (e.g., Vandemeent et al., 2016). In the problems we have considered, the optimal policy is represented simply in terms of which action to take for each state of the MDP. However, we might expect that people have a more systematic representation of these strategies, such as "in situations with this feature, do this," "repeat this step this many times," and "while you are uncertain, do this." These representations correspond to the kinds of control structures commonly used in computer programming, such as if-then statements and loops. In seeking to understand human cognition, we could thus potentially benefit from searching a space of metacognitive policies that are specified in the form of simple programs. This remains a challenging problem, but it is one where there is already promising initial progress.

9.2.3   Understanding the behavior of groups and organizations

If anyone were to be called the founder of the field of bounded rationality, it would be Herbert Simon. He was a political scientist who developed the foundational concepts of bounded rationality by studying decision-making in organizations (Simon, 2013). The theory of resource rationality is, in many ways, a modern, more precise version of Simon's theory of bounded rationality. So far, we have formalized and tested Simon's ideas in research on individual decision-making. To come full circle, future work should extend our efforts to formalize Simon's idea of bounded rationality to capture the intricacies of organizational decision-making that motivated that idea. Looking even further ahead, we hope that resource-rational analysis will also be applied to understanding and improving societal decision-making (see Section 8.8.3).

Organizational decision-making and societal decision-making are instances of **collective decision-making**. Collective decision-making can be understood as a cognitive process distributed across multiple interacting minds. The cognitive architecture of the collective can be modeled as a collection of independent modules, one for each individual, connected by capacity-limited channels of communication (Vélez et al., 2023). The limited cognitive resources of the individual minds and the limited capacity of the communication channels that connect them are serious impediments to optimal decision-making. Resource-rational analysis can help us understand and mitigate the implications of these constraints. Developing resource-rational models of collective decision-making is thus a promising approach to better understanding the goals, cognitive strategies, representations, and limitations of decision-making in groups, organizations, and society.

More generally, resource-rational models of collective decision-making could be used to ground the analysis of the collective behavior of groups, organizations, and societies on fundamental research on the mind of the individual. In this way, resource-rational models could become a bridge between ideas from cognitive psychology and neuroscience to those from the social sciences, including economics, organization science, sociology, history, and political science. Moreover, the resource-rational framework can be applied to optimally design interventions for improving organizational and societal decision-making. These applications could confer significant benefits to organizations and society (see Section 8.8.3).

9.3 A new view of the mind

The value of understanding rational action is twofold: it gives us a way to make predictions about how people will behave, and it provides a guide for helping people to make better decisions. By offering a different perspective on how to think about rationality—and a different way to construct models of human behavior—the approach we have presented in this book has the potential to guide both prediction and intervention in new directions. This new view of the mind is relevant to research in psychology and cognitive science. Still, perhaps its greatest value is in providing a way for insights from these fields to be incorporated into other disciplines.

Most theories used in economics, management, organization science, sociology, political science, other social sciences, and the humanities have to make assumptions about the human mind. In many cases, these assumptions are crucial. The rise of behavioral economics illustrates how updating our assumptions about the human mind can allow these fields to make rapid progress (Thaler, 2015). The fact that these updates occurred decades after heuristics and biases had been discovered in psychology highlights that the rate at which findings about the human mind are translated into other fields can be improved. This translation is, in fact, so slow and imperfect that the assumptions of many existing theories about human behavior are outdated, inaccurate, or simplistic.

One way to facilitate the transfer of knowledge from psychology and cognitive science to other disciplines is to provide those fields with quantitative models of human behavior that provide a reasonably up-to-date summary of what is known about the human mind. We see resource-rational models of human decision-making as a promising format for facilitating such a process. Indeed, economists have already begun to develop and use resource-rational models of decision-making (e.g., Gabaix, 2023). This approach could also benefit other social sciences and the decision-makers who rely on them. This, in turn, could inform the policies and interventions designed based on such models. We hope that, by facilitating this transfer of knowledge, resource-rational analysis will help society realize the Enlightenment ideal that our society should base its decisions on reason and the best available evidence.

10

Appendix: Mathematical Details

This appendix provides a guide to notation as well as a formal description of the metalevel MDP framework and various techniques for solving metalevel MDPs. The formalization presented here is an adaptation of the framework proposed in Hay (2016) for the purpose of cognitive modeling.

10.1 Notation

Object	Notation
Expected value of $f(x)$ given y	$\mathbb{E}_{x\|y}[f(x)]$
Probability of a random variable taking on a discrete value x	$p(x)$
Probability density at x	$p(x)$
Set	$\mathcal{A} = \{1, 2, 3\}$
x is a member of the set $\mathcal{A}$	$x \in \mathcal{A}$
x is not a member of the set $\mathcal{A}$	$x \notin \mathcal{A}$
Normal distribution with mean μ and standard deviation σ	$\mathcal{N}(\mu, \sigma^2)$
The set of all distributions over the set $\mathcal{X}$	$\Delta(\mathcal{X})$
The argument x for which a function $f(x)$ takes its highest possible value	$\arg\max_x f(x)$
The argument x for which a function $f(x)$ takes its lowest possible value	$\arg\min_x f(x)$
The best possible value of i	i^* or $i^\star$
The set of all possible pairs (a, b) of elements a from set $\mathcal{A}$ and elements b from set $\mathcal{B}$	$\mathcal{A} \times \mathcal{B}$

10.2 Markov decision processes

A metalevel Markov decision process is an extension of a standard Markov decision process (MDP), illustrated in Figure 10.1. Thus, we begin with a

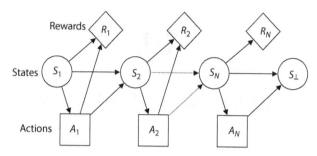

FIGURE 10.1. **Markov decision processes.** A Markov decision process (MDP) formalizes the problem of acting adaptively in a dynamic environment. The agent executes actions (squares) that change the state of the world (circles) and generate rewards (diamonds), which the agent seeks to maximize. The arrows indicate direct causality; thus, the reward and state at each time step depend only on the state and action at the previous time step, and the agent selects an action based only on the current state. The dotted arrow indicates the elided sequence of states between the first two and the last two.

brief overview of standard MDPs. See Puterman (2014) and Sutton and Barto (2018) for more thorough overviews.

MDPs are the standard formalism for modeling the sequential interaction between an agent and a stochastic environment. An MDP is defined by a set of states $\mathcal{S}$, a set of actions $\mathcal{A}$, a transition function T, and a reward function R. A state $s \in \mathcal{S}$ specifies the relevant state of the world. An action $a \in \mathcal{A}$ is an action the agent can perform. The transition function $T : \mathcal{S} \times \mathcal{A} \rightarrow \Delta(\mathcal{S})$[1] encodes the dynamics of the world as a distribution of possible future states for each possible previous state and action. Finally, the reward function $R : \mathcal{S} \times \mathcal{A} \rightarrow \mathbb{R}$ specifies the expected[2] reward or utility for executing a given action in a given state. We additionally assume an initial state s_1 that the environment is initialized in and a set of terminal states $\mathcal{S}_\perp$ such that the episode ends when the agent reaches one of those states. An *episode* describes one interaction between the agent and the environment (beginning in the initial state and ending in a terminal state).

---

1. One could equally well specify a stochastic reward function $R : \mathcal{S} \times \mathcal{A} \rightarrow \Delta(\mathbb{R})$. Because the optimal policy only depends on the expected reward, we do not consider this case here. However, a stochastic reward function would affect the performance of a learning algorithm; this could easily be integrated into the framework.

2. $\Delta(\mathcal{S})$ denotes the set of all distributions over the set $\mathcal{S}$. Note that this definition is equivalent to defining the transition function as a probability mass function (i.e., $T : \mathcal{S} \times \mathcal{A} \times \mathcal{S} \rightarrow [0, 1]$). We will use $T(s' \mid s, a)$ to denote the probability of transitioning to state s' when executing action a in state s.

10.2.1 Optimal policies and value functions

The solution to an MDP is a policy $\pi : \mathcal{S} \rightarrow \Delta(\mathcal{A})$ that selects which action to perform next given the current state. That is, $a_t \sim \pi(s_t)$. The goal is to find a policy that maximizes the expected cumulative reward attained, that is, the *return*. The optimal policy is thus defined

$$\pi^* = \arg\max_{\pi} \mathbb{E}\left[\sum_{t=1}^{N} R(s_t, a_t) \,\middle|\, a_t \sim \pi(s_t)\right], \tag{10.1}$$

where N is the time step at which the episode terminates (when $s_{t+1} \in \mathcal{S}_\perp$). Note that the expectation implicitly conditions on the transition function, i.e., $s_{t+1} \sim T(s_t, a_t)$.

How can we identify such a policy? This question is the subject of a huge field of research in articifical intelligence, and countless methods have been developed to answer it. Many of these methods draw on the concept of a *value function*. The *state* value function (or just "value function") is defined as

$$V^\pi(s) = \mathbb{E}\left[\sum_{t=1}^{N} R(s_t, a_t) \,\middle|\, s_1 = s, \; a_t \sim \pi(s_t)\right]. \tag{10.2}$$

It specifies the expected total reward one will receive if one begins in state s and selects actions according to the policy π. Similarly, the *action* value function (or "state-action value function") is defined as

$$Q^\pi(s, a) = \mathbb{E}\left[\sum_{t=1}^{N} R(s_t, a_t) \,\middle|\, s_1 = s, \; a_1 = a, \; a_{t \neq 1} \sim \pi(s_t)\right]. \tag{10.3}$$

The action value function is just like the state value function except that it also specifies the first action to be taken.

The value functions for the optimal policy are called the optimal value functions. They can be defined simply as $V^* = V^{\pi^*}$ and $Q^* = Q^{\pi^*}$. By combining Equations 10.2 and 10.3 with Equation 10.1, we can see that the optimal value functions specify the maximal expected reward one could expect to gain beginning with a given state (and action) under any policy:

$$V^*(s) = \max_{\pi} \mathbb{E}\left[\sum_{t=1}^{N} R(s_t, a_t) \,\middle|\, s_1 = s, \; a_t \sim \pi(s_t)\right],$$

$$\tag{10.4}$$

$$Q^*(s, a) = \max_{\pi} \mathbb{E}\left[\sum_{t=1}^{N} R(s_t, a_t) \,\middle|\, s_1 = s, \; a_1 = a, \; a_{t \neq 1} \sim \pi(s_t)\right].$$

Putting aside for now the problem of identifying these functions, the optimal policy can be defined as simply

$$\pi^*(s) = \text{Uniform}\left(\arg\max_a Q^*(s, a)\right). \qquad (10.5)$$

That is, the optimal policy selects the action that produces the greatest expected long-term reward, breaking ties randomly.[3] When modeling human data, one typically assumes that this maximization is performed imperfectly. In particular, we will assume that actions are drawn according to a softmax function (or a Boltzmann distribution),

$$\pi(a \mid s) \propto e^{\beta \cdot Q^*(s,a)}, \qquad (10.6)$$

where the inverse-temperature parameter β controls how well the policy maximizes, behaving completely randomly when $\beta = 0$ and approaching the optimal policy as $\beta \to \infty$.

This concludes our brief overview of MDPs. We are now ready to describe metalevel MDPs.

10.3   Metalevel Markov decision processes

Metalevel Markov decision processes (metalevel MDPs) extend the standard MDP formalism to model the sequential decision problem posed by resource-bounded computation (Hay et al., 2012). Like a standard MDP, a metalevel MDP is defined by sets of states and actions, and transition and reward functions. However, as illustrated in Figure 10.2, the states correspond to mental states and the actions correspond to computations (cognitive operations). The metalevel transition function describes how computations update mental states. The metalevel reward function captures both the internal costs (e.g., time) and external benefits (e.g., better decisions) of computation.

Formally, we define a metalevel MDP by a set of mental states $\mathcal{M}$, a set of computations $\mathcal{C}$, a transition function T, and a reward function R. These four components are analogous to the states, actions, transition function, and reward function in a standard MDP. We additionally define a set of world states $\mathcal{W}$, upon which the transition and reward functions depend. This is the only formal distinction between a metalevel MDP and a standard MDP. However, as we discuss in Section 10.7, we can sometimes convert metalevel MDPs into

---

3. To be more precise, Equation 10.5 defines the *maximum entropy* optimal policy, that is, the optimal policy whose action distributions are maximally random. For simplicity, we will refer to the maximum entropy optimal policy as simply "the" optimal policy.

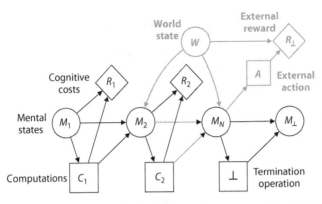

FIGURE 10.2. **Metalevel Markov decision processes**. A metalevel MDP formalizes the problem of thinking efficiently within one's own internal mental environment. The agent executes computations that update their mental state and incur cognitive cost. When the agent executes the termination operation $\perp$, they take an action in the external environment and receive an external reward. The elements that capture the external environment are indicated in bold. These formally distinguish metalevel MDPs from standard MDPs.

equivalent MDPs by marginalizing over the world state.[4] We now describe the five components of a metalevel MDP in more detail.

To make things concrete, we will use a simple running example based on the tallying heuristic for deciding between two possible actions (Gigerenzer and Gaissmaier, 2011). We will consider applying this heuristic to the decision of which car to purchase. To follow this heuristic, we consider a sequence of "cues" that might discriminate between the options, marking which car each cue favors (which has better gas mileage? which is more comfortable? etc.). As we go, we simply count up how many cues have favored each car. Then, after considering some number of cues, we pick the car that is favored by more cues. But how many cues should we consider? Gigerenzer and Gaissmaier propose considering a fixed number of cues (and, in the case of a tie, considering more cues until one car has more in its favor). Is this the best version of tallying we could use? To answer this question, we can define tallying as a metalevel MDP.

A well-documented and (hopefully) beginner-friendly Julia implementation of the tallying metalevel MDP can be found at https://github.com/fredcallaway/metamdp-example.

---

4. This is similar to the *belief MDP* transformation of a partially observable MDP (POMDP; Kaelbling et al., 1998). See Section 10.6.4 for additional discussion on the relation between metalevel MDPs and POMDPs.

10.3.1   World states

A world state $w \in \mathcal{W}$ captures the state of the world that is relevant to the agent's current task. Importantly, the agent does not have direct access to the world state, but must infer it from the outcome of computations it performs. Formally, the world state includes any information that is not known to the agent, but affects either the reward or transition functions. In the tallying example, the world state specifies the ratio of cues that favors one car versus the other; thus, $\mathcal{W}_{\text{tally}} = [0, 1]$.

10.3.2   Mental states

A mental state $m \in \mathcal{M}$ captures the agent's internal state, as relevant to the task at hand. The interpretation of a mental state can vary from model to model. In some cases, it will correspond to a belief or a representation of the world, but it can also capture arbitrary variables in a cognitive model. In the tallying example, the mental state specifies the number of cues one has considered that favor each option. Formally, we define each $m \in \mathcal{M}_{\text{tally}}$ as a tuple (x, y), where x is the number of cues favoring car X and y is the number of cues favoring car Y.

Analogously to standard MDPs, we additionally specify an initial mental state, m_1. This is the mental state the agent has at the beginning of each task instance. In the tallying example, the initial mental state is defined $m_1 = (0, 0)$. Additionally, all metalevel MDPs have a single terminal state $m_\perp$, which is only entered when computation is terminated (as described below).

10.3.3   Computations

A computational operation $c \in \mathcal{C}$ is a primitive operation afforded by the agent's cognitive architecture. Formally, it is a metalevel action that changes the mental state in much the same way as an external action might change the world state. In a metalevel MDP model, all cognition can be broken down into a sequence of these computations, but the model makes no attempt to explain how those basic operations are themselves implemented. The concept is thus very similar to *elementary information processes* (Chase, 1978; Simon, 1979; Posner and McLeod, 1982; Payne et al., 1988). In the tallying example, there is a single computation, which corresponds to considering another cue.

All metalevel MDPs include a special computation, the termination operation $\perp$, which indicates that computation should be terminated. Upon termination, the agent performs an external action, specifically the action that has maximal expected utility given the current mental state. Thus, the most fundamental metalevel problem—how long to compute—is captured by the decision about when to execute $\bot$. In the tallying example, executing $\bot$ corresponds to purchasing the car that is favored by more cues (choosing randomly in the case of a tie.)

10.3.4 Transition function

The transition function $T : \mathcal{M} \times \mathcal{C} \times \mathcal{W} \to \Delta(\mathcal{M})$ describes how computation updates mental states. Formally, $T(m, c, w)$ is a distribution of possible new mental states that would result from performing a computation c in mental state m when the true state of the world is w. At each time step, the next mental state is sampled from this distribution:

$$m_{t+1} \sim T(m_t, c_t, w). \tag{10.7}$$

Terminating computation (executing $\bot$) always transitions to the terminal state, $m_\bot$:

$$T(m, \bot, w) = \mathrm{Uniform}(\{m_\bot\}). \tag{10.8}$$

In the tallying example, the transition function specifies the probability that each cue count will be incremented when one considers another cue:

$$T_{\mathrm{tally}}(m_t, c_t, w) = \begin{cases} (x_t + 1, y_t) & \text{with probability } w \\ (x_t, y_t + 1) & \text{with probability } 1 - w, \end{cases} \tag{10.9}$$

where $m_t = (x_t, y_t)$.

10.3.5 Reward function

The metalevel reward function $R : \mathcal{M} \times \mathcal{C} \times \mathcal{W} \to \mathbb{R}$ describes both the costs and benefits of computation. It is defined in terms of three components: a cost function that specifies the cost of executing each computation, an action policy act that selects an action to take given a mental state, and a utility function U that specifies the utility of each possible action in each world state.

The cost of computation is captured in the reward for nonterminal operations:

$$R(m, c, w) = -\mathrm{cost}(m, c) \text{ for } c \neq \bot. \tag{10.10}$$

We assume that the cost of a computation can depend on the current mental state but not on the state of the world. The cost of computation may include multiple factors. At a minimum, it captures the opportunity cost of the time spent executing the computation (rather than taking actions in the world). The simplest choice is to assume a constant cost for each computation executed. This is a natural choice for the tallying example, with total cost proportional to the number of cues considered.

The benefits of computation are captured by the reward for the termination operation, $R(m, \perp, w)$. Intuitively, the benefit of computation is that it allows one to take better actions in the world. Formally, the reward for termination is defined as the utility of the external action that the agent would execute given the current mental state,

$$R(m, \perp, w) = U\big(w, \text{act}(m)\big), \tag{10.11}$$

where $U(w, a)$ specifies the utility of executing action a in the world state w and act is the *action selection policy*, which chooses an action to take based on the current mental state. To simplify notation, we will assume that act selects an action deterministically, but it can also return a distribution over actions. This policy should be very simple. That is, the mental state should contain sufficient information to choose an action without much additional computation.

In the tallying example, act deterministically selects the car with more favorable cues, selecting randomly otherwise. Specifying U is less straightforward. For simplicity, we assume that the utility derived from car X is proportional to the ratio of cues favoring X. This results in

$$U_{\text{tally}}(w, a) = \begin{cases} w & \text{if } a = X \\ 1 - w & \text{if } a = Y. \end{cases} \tag{10.12}$$

Substituting our definitions of U and act into Equation 10.11 yields:

$$R_{\text{tally}}(m, \perp, w) = \begin{cases} w & \text{if } x_t > y_t \\ 1 - w & \text{if } x_t < y_t \\ 1/2 & \text{if } x_t = y_t, \end{cases} \tag{10.13}$$

where the $1/2$ comes from taking the average of w and $1 - w$.

Equation 10.11 may appear to constrain us to simple one-shot decision-making tasks, such as the ones tallying is intended for. However, it is not as restrictive as it first appears. For example, in Section 6.2.4, we model a memory recall task by assuming that act can only perform the "recall" action when the activation of a memory exceeds a threshold. In Section 6.2.3, we model a planning task by defining a abstractly, as a sequence of concrete actions (or, more generally, an *option*; Sutton et al., 1999).[5]

---

5. This does require that all computation is executed before any external action is performed, but this constraint does not reduce performance in deterministic environments.

10.4   Metalevel policies

If a metalevel MDP defines the problem a cognitive process must solve, a metalevel policy defines the solution. It is a strategy for selecting which cognitive operation to execute next given the current mental state. Formally, the metalevel policy $\pi : \mathcal{M} \to \Delta(\mathcal{C})$ is a mapping from beliefs to distributions over computations. At each time step, the next computation is drawn from this distribution: $c_t \sim \pi(m_t)$.

How should we determine this policy? The classical cognitive modeling approach is to specify a plausible strategy, perhaps motivated by aspects of human behavior. In Section 6.2.3, we show how classical heuristics for decision-tree search can be naturally modeled as policies in a metalevel MDP. However, in the resource-rational approach pursued here, we take a different approach. Specifically, we are interested in the *optimal policy* for the metalevel MDP. Paralleling Equation 10.1, the optimal metalevel policy is defined as

$$\pi^* = \arg\max_{\pi} \mathbb{E}\left[\sum_{t=1}^{N} R(m_t, c_t, w) \,\middle|\, c_t \sim \pi(m_t) \right]. \tag{10.14}$$

That is, it maximizes the expected return. In a metalevel MDP the return can be broken down into two components, capturing the expected utility of the external action and the cost of the computations necessary to choose that action, respectively:

$$\pi^* = \arg\max_{\pi} \mathbb{E}\left[U\big(w, \mathrm{act}(m_N)\big) - \sum_{t=1}^{N-1} \mathrm{cost}(m_t, c_t) \,\middle|\, c_t \sim \pi(m_t) \right]. \tag{10.15}$$

This is the definition of optimal sequential cognitive processes given in Equation 10.38.[6] It emphasizes that the optimal policy is the cognitive strategy that best trades off between the costs and benefits of computation.

The optimal policy for the tallying example is illustrated in Figure 10.3. When the cost of computation is high relative to the stakes of the decision, we see that the optimal tallying policy considers cues until either one option has a lead of two cues in its favor, or thirteen cues have been considered in total. This is interestingly different from the suggestion of Gigerenzer and Gaissmaier (2011), to first consider a fixed number of cues and then make a choice

---

6. Note that the expectation implicitly depends on the transition function, which was explicit in Equation 10.38, as well as the initial mental state m_1 and the set of possible computations.

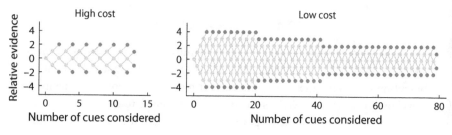

FIGURE 10.3. **Optimal policies for the tallying metalevel MDP.** The policy is illustrated as a graph, where each node represents a mental state. States in which the optimal policy continues computing are light gray, and states in which the optimal policy terminates are dark gray. The two panels show the optimal policy under different levels of computational cost (0.01 and 0.002, in units defined by the maximal action utility of 1). The code that generated this figure can be found at https://github.com/fredcallaway/metamdp-example.

as soon as one leads by any amount. When deliberation is less costly (or the decision is more important), one option has to lead by four cues to be chosen; but over time, this requirement becomes less strict, with a decision being made after considering at most 79 cues. This resembles the "collapsing boundaries" that are often found in optimal evidence accumulation models (Drugowitsch et al., 2012).

Unfortunately, identifying optimal metalevel policies is substantially more challenging than writing down their definition. In Section 10.8, we discuss various strategies for tackling this problem.

10.5   Two views of computation

So far, we have characterized computation as a process of executing mental actions that update an agent's internal state. These mental actions can be contrasted with external actions that update the state of the environment—but they are fundamentally the same type of thing: *actions* In the language of Nicholas Hay (2016; Chapter 7), this is the *mechanical view* of computation. This view is nonrestrictive; it is relatively easy to see how any algorithm or cognitive model one might concoct could be formalized in this way.

An alternative view, the *Bayesian view*, takes a stronger stance on the nature of computation. It formalizes computations as experiments that generate information about the world (Matheson, 1968). The results of these experiments are synthesized, by Bayesian inference, into posterior distributions over the utility of different possible external actions, thus informing the agent's choice of which action to take.

It is worth emphasizing that this view is quite different from the standard Bayesian approach in cognitive science (e.g., Tenenbaum et al., 2011). While the standard approach treats cognition as a problem of drawing inferences from data, this view treats cognition as a problem of generating the data that drives inference. A closer analogue in cognitive science is found in models of active learning (Gureckis and Markant, 2012; Gottlieb et al., 2013), but in that case, the data is assumed to be found in the world rather than generated in the mind.

Importantly, the two views are not mutually exclusive. They are often compatible interpretations of a single system. The mechanical model emphasizes the process of computation, while the Bayesian model emphasizes its function. In this way, the mechanical and Bayesian models are analogous to Marr's algorithmic and computational levels. Unlike in Marr's levels, however, adopting the Bayesian model is more than just an interpretation; it has practical consequences for what one can do with the model. Specifically, it puts constraints on the types of computation one can consider. All Bayesian metalevel MDPs have a mechanical interpretation, but not vice versa. On the other hand, adopting the Bayesian view has conceptual and technical advantages. Specifically, it provides a formal link between mental states and world states, and it allows us to convert metalevel MDPs into standard MDPs.

In the following section, we define Bayesian metalevel MDPs as a special case of the general metalevel MDP framework outlined above.

10.6 Bayesian metalevel MDPs

Taking the Bayesian view amounts to putting a restriction on the set of mental states and the transition function. Specifically, the mental states must correspond to posterior distributions over the state, and the transition function must encode and describe a Bayesian updating procedure. We detail these two requirements below.

10.6.1 Belief states

When adopting the Bayesian view, mental states correspond to *beliefs*, formally expressed as distributions over the world state. We will use the notation b_m to denote the Bayesian belief associated with mental state m. When it is clear from context, we will drop the m subscript, e.g., using b_t in place of b_{m_t}.

Mapping mental states to Bayesian beliefs is powerful because it provides a formal link between mental states and world states. For example, we can denote the probability of a world state under a belief state as

$b_m(w) = \Pr(W = w \mid M = m).$[7] Similarly, we can express expectations about functions of world state given the mental state as $\mathbb{E}_{w \sim b_m}[f(w)]$. This allows us to, for example, define an action policy that takes the optimal action given the current mental state;

$$\text{act}(m) = \text{Uniform}\left(\underset{a}{\arg\max}\ \underset{w \sim b_m}{\mathbb{E}}[U(w, a)]\right). \qquad (10.16)$$

To be more precise, this policy randomly selects one of the actions that has maximal expected utility given the mental state. This is the default action policy in a Bayesian metalevel MDP, meaning that it is one less choice we have to make as cognitive modelers.

In the Bayesian view, the intial mental state m_1 has a special interpretation. It is the *prior*, the distribution over the world state the agent assumes before performing any computation. By default, we assume that this prior is accurate, i.e., $b_1(w) = \Pr(W = w)$. However, in the fixation study described in Section 6.2.2, we had to assume some bias in the prior to fully capture human behavior.

10.6.2 Bayesian updating

In the Bayseian view, computations correspond to experiments that generate information about the world state; this information is then integrated into a belief by Bayesian inference. The transition function describes this process. Formally, each computation defines a state-dependent distribution of computational "outcomes" $p_c(o \mid w)$. Importantly, note that this is the outcome of a computation, rather than an external action (as we used the term in Chapter 3). Given the previous mental state m_t and the outcome o_t, the new mental state m_{t+1} is defined such that

$$b_{t+1}(w) = p(w \mid m_t, o_t) = \frac{p_c(o_t \mid w) b_t(w)}{p(o_t \mid m_t, c_t)}, \qquad (10.17)$$

where the second equality is the application of Bayes rule, updating the prior b_t given the likelihood $p_c(o \mid w)$. The transition function describes the full process of sampling an observation and updating the belief accordingly. Denoting the update in Equation 10.17 as "bayes-update," the full transition function is defined as

$$o_t \sim p_c(\cdot \mid w),$$
$$m_{t+1} = \text{bayes-update}(m_t, o_t, p_c). \qquad (10.18)$$

---

7. For notational convenience, we assume in this chapter that $\mathcal{W}$ is countable. The definitions can easily be extended to the continous case.

10.6.3   Tallying as a Bayesian metalevel MDP

As emphasized earlier, the Bayesian view is not incompatible with the mechanical view. All Bayesian metalevel MDPs have a mechanical view (or at least, all those that can be implemented on a physical machine). And in some especially fortuitous cases, one may even find that a model one initially specified in mechanical terms has a Bayesian interpretation as well. As luck would have it, our tallying example is just such a case! Indeed, it is a minor modification of one of the first explicitly formalized metalevel MDPs (Hay et al., 2012).[8]

Viewing tallying from a Bayesian perspective, the mental state corresponds to a distribution over the ratio of cues in favor of each car, w. The standard choice for a distribution over ratios and probabilities is the beta distribution; thus, we define $b_t = \text{Beta}(\alpha_t, \beta_t)$. The initial values, α_1 and β_1, specify a prior over w. A natural choice is $\alpha_1 = \beta_1 = 1$, which results in a uniform distribution over the range $[0, 1]$.

Turning to the transition function, recall that each computation corresponds to considering an additional cue, which may be in favor of one car or the other. We can think of which car the cue favors as the outcome of the computation, o. By definition, the probability that each cue favors car X is w, and so we have $o_t \sim \text{Bernoulli}(w)$. Finally, we must specify how these observations are integrated into the belief state ("bayes-update" in Equation 10.18). Because the beta distribution is the conjugate prior for the Bernoulli distribution, this update has a very simple form:

$$\text{bayes-update}(m_t, o_t, p_c) = \begin{cases} \text{Beta}(\alpha_t + 1, \beta_t) & \text{if } o_t = 1 \\ \text{Beta}(\alpha_t, \beta_t + 1) & \text{if } o_t = 0. \end{cases} \tag{10.19}$$

Not only is this form simple, it is remarkably similar to Equation 10.9. Indeed, given that $p(o_t = 1) = w$, they are equivalent. The only difference between α_t and x_t (and between β_t and y_t) is that the x and y are assumed to be initialized to 0, while α and β are initialized to 1 (or some other values to capture a nonuniform prior). That is, $b_t = \text{Beta}(\alpha_1 + x_t, \beta_1 + y_t)$, providing a direct link between the Bayesian belief state and the mechanical mental state.

---

8. Specifically, it corresponds to the one-armed Bernoulli metalevel probability model. The modification is in the termination reward. We assume that the agent can choose between two options with value w and $1 - w$, whereas Hay et al. assume a choice between w and a known constant value. It should go without saying that luck had nothing to with this correspondence.

10.6.4   Relation to partially observable Markov decision processes

For those who are familiar with the reinforcement learning and planning literatures, the equations above will likely look familiar. Specifically, they closely resemble the equations associated with belief updating in a partially observable Markov decision process (POMDP; Kaelbling et al., 1998). POMDPs are generalizations of MDPs where the agent does not know the state, but instead receives an observation conditional on the state and action at each time step: $o_t \sim O(s_{t+1}, a_t)$. Given these observations, the agent maintains a belief about the current state using a Bayesian update similar to Equation 10.17, but additionally accounting for the possibility that the state changes:

$$b_{t+1}(s_{t+1}) = p(s_{t+1} \mid b_t, o_t) = \frac{O(o_t \mid s_{t+1}, a_t) \sum\limits_{s_t \in \mathcal{S}} T(s_{t+1} \mid s_t, a_t) b_t(s_t)}{p(o_t \mid b_t, a_t)}. \tag{10.20}$$

Bayesian metalevel MDPs can thus be understood as a subset of POMDPs in which the state never changes. In this case, the belief update reduces to

$$b_{t+1}(s) = \frac{O(o_t \mid s, a_t) b_t(s)}{p(o_t \mid b_t, a_t)}, \tag{10.21}$$

which is exactly analogous to Equation 10.17. Metalevel MDPs additionally require that all but one action yield strictly negative reward and that the remaining action, $\perp$, leads to a terminal state.

Given that Bayesian metalevel MDPs are a special case of POMDPs, one could reasonably ask why we should bother with metalevel MDPs at all. Indeed, other researchers have proposed using POMDPs to model an agent's internal environment (e.g., Acharya et al., 2017; Hoppe and Rothkopf, 2019; Chen et al., 2021; Oulasvirta et al., 2022). However, there are conceptual and technical advantages to using the more restrictive formalism of metalevel MDPs. Conceptually, it is useful to formally distinguish between mental states and world states, and between computational actions and external actions. Although it is sometimes natural to map the belief state in a POMDP to a mental state, a mental state cannot always be reduced to a belief (i.e., when we are not taking a strictly Bayesian view of computation). From a technical perspective, POMDPs describe a very general and challenging class of problems; as a result, general-purpose POMDP solvers are notoriously inefficient.[9]

---

9. "The best thing about POMDPs is that everything's a POMDP. But the worst thing about POMDPs is that everything's a POMDP." (Michael Littman, personal communication).

Focusing on the specific case defined by metalevel MDPs allows us to develop more targeted and efficient solution strategies (e.g., Section 10.9).

Nevertheless, noting the formal similarities between Bayesian metalevel MDPs and POMDPs allows us to take advantage of a powerful tool from the POMDP literature. Specifically, we can draw on the concept of *belief MDPs* (Kaelbling et al., 1998) to convert a Bayesian metalevel MDP into an equivalent MDP, where the world state has been marginalized out. We refer to this as the "marginalized" metalevel MDP.[10]

10.7   Marginalized metalevel MDPs

As observed by Kaelbling et al. (1998), it is possible to convert a POMDP into an MDP by defining modified transition and reward functions that take beliefs (rather than states) as input and return expected transition probabilities and rewards, marginalizing over the state of the world. This is desirable because it allows one to identify the optimal policy for a POMDP using tools developed to solve MDPs. As outlined below, we can apply a similar strategy to metalevel MDPs. Specifically, we define *marginal* versions of the reward and transition functions that do not depend on the world state. Together with the set of mental states $\mathcal{M}$ and computational "actions" $\mathcal{C}$, this yields a standard MDP. We define the marginalized transition and reward functions below.

10.7.1   Marginal transition function

The marginal transition function $T : \mathcal{M} \times \mathcal{C} \to \Delta(\mathcal{M})$ is most easily defined in generative form:

$$w \sim b_t,$$
$$m_{t+1} \sim T(m_t, c_t, w). \tag{10.22}$$

One simply samples the world state from the belief before applying the transition dynamics. This is sufficient to sample from $T(m, c)$. However, we will often need an explicit probability mass function. This is defined as

$$T(m_{t+1} \mid m_t, c_t) = \mathop{\mathbb{E}}_{w \sim b_t} \left[T(m_{t+1} \mid m_t, c_t, w) \right]. \tag{10.23}$$

This expression cannot be simplified in the general case. In practice, we will work with conjugate or discrete beliefs that make this integration tractable. The general strategy is to derive a posterior predictive distribution over the

---

10. We don't just call them "belief metalevel MDPs" because the state still corresponds to a mental state, which can function as more than just a belief state (see Section 10.7.3).

observation, $p(o_t \mid c, m)$, which can be transformed into the transition function by applying bayes-update to the support of this distribution.

In the tallying example, the relevant posterior predictive is given by a Beta-Bernoulli process; the marginal transition is thus defined as

$$T_{\text{tally}}(m_{t+1} \mid m_t, c_t) = \begin{cases} \frac{\alpha_t}{\alpha_t + \beta_t} & \text{if } m_{t+1} = (\alpha_t + 1, \beta_t) \\ \frac{\beta_t}{\alpha_t + \beta_t} & \text{if } m_{t+1} = (\alpha_t, \beta_t + 1). \end{cases} \tag{10.24}$$

10.7.2 Marginal reward function

The marginal reward function $R : \mathcal{M} \times \mathcal{C} \to \mathbb{R}$ is defined as

$$R(m, c) = \underset{w \sim b_m}{\mathbb{E}} [R(m, c, w)]. \tag{10.25}$$

For $c \neq \perp$, $R(m, c, w)$ does not depend on w; we thus have:

$$R(m, c) = -\text{cost}(m, c). \tag{10.26}$$

The reward for terminating, however, may depend on the state of the world; we must marginalize it out. This results in:

$$R(m, \perp) = \max_a \underset{w \sim b_m}{\mathbb{E}} [U(w, a)]. \tag{10.27}$$

That is, the marginal termination reward is simply the maximal expected utility of any action. Although this may seem counterintuitive, there is a simple intuition: if you select an action that has maximal expected utility, the expected utility of the chosen action will be maximal. A skeptical reader might ask whether this is an equivocation: the first use of "expected" refers to our subjective predictions, while the latter use refers to the actual, objective outcome. However, because we assume beliefs to be accurate (Equation 10.30), the subjective and objective expectations are one and the same. This can be seen in the following derivation:

$$
\begin{aligned}
R(m, \perp) &= \underset{w \sim b_m}{\mathbb{E}} [R(m, \perp, w)] \\
&= \underset{w \sim b_m}{\mathbb{E}} [U(w, \text{act}(m))] \\
&= \underset{w \sim b_m}{\mathbb{E}} \left[U\left(w, \arg\max_a \underset{w' \sim b_m}{\mathbb{E}} [U(w', a)]\right)\right] \\
&= \max_a \underset{w \sim b_m}{\mathbb{E}} [U(w, a)].
\end{aligned}
\tag{10.28}
$$

The final line follows from $f\left(\arg\max_a f(a)\right) = \max_a f(a)$, where $f(a) = \mathbb{E}_{w \sim b_m}\left[U(w, a)\right]$. Note that the logic generalizes to the case where $\text{act}(m)$ samples from the set of optimal actions.

In the tallying example, the marginal termination reward can be derived as

$$R_{\text{tally}}(m, \perp) = \max_a \mathbb{E}_{w \sim b_m}\left[U(w, a)\right]$$

$$= \max\left\{\mathbb{E}_{w \sim b_m}\left[U(w, X)\right],\ \mathbb{E}_{w \sim b_m}\left[U(w, Y)\right]\right\}$$

$$= \max\left\{\mathbb{E}_{w \sim b_m}\left[w\right],\ \mathbb{E}_{w \sim b_m}\left[1 - w\right]\right\} \tag{10.29}$$

$$= \max\left\{\frac{\alpha}{\alpha + \beta},\ \frac{\beta}{\alpha + \beta}\right\},$$

where $\frac{\alpha}{\alpha+\beta}$ is the posterior mean of the Beta distribution b_m.

10.7.3   Marginalizing mechanical metalevel MDPs

Being able to marginalize a metalevel MDP model is very desirable, as it allows us to use general MDP-solving techniques to identify optimal metalevel policies (Section 10.8). But if marginalization is only possible for Bayesian metalevel MDPs, this would limit our ability to use the more general mechanical definition (Section 10.3). Fortunately, marginalization is sometimes still possible when we are not taking the Bayesian view. Inspecting Equations 10.23 and 10.27, we see that computing the marginalized transition and reward functions simply requires taking an expectation over the world state with respect to the belief state. Thus, as long as we can define a belief state associated with any mental state, we can apply the marginalization. However, for the marginalization to result in a valid MDP that is truly equivalent to the original metalevel MDP, the belief state must satisfy two conditions: it must be *accurate* and *complete*.

The accuracy requirement is relatively straightforward. It simply states that the probability the belief state b_m assigns to a world state w is the actual probability that the world is in state w given that you arrived in mental state m. Formally, accuracy requires that

$$b_t(w) = p(w \mid m_t) = \Pr(W = w \mid M_t = m_t). \tag{10.30}$$

It is easy to see why this property is necessary. If one computes the marginal transition and reward functions given incorrect assumptions about the world state, those functions will also be incorrect. In principle, the accuracy requirement does not impose any retrictions on the metalevel MDP. It can always be satisfied by defining the belief state using Bayes' rule,

$$b_t(w) = \frac{p(m_t \mid w)p(w)}{p(m_t)}. \tag{10.31}$$

In practice, however, one must specify the mental state in such a way that these probabilities can be computed very quickly.

The completness requirement is more nuanced, and restrictive. It requires that the belief state contain all the information about the world state that it could possibly have, given the full history of the episode up to that time point. Formally, completeness requires that

$$b_t(w) = p(w \mid \boldsymbol{m}_{1:t}, \boldsymbol{c}_{1:t}). \tag{10.32}$$

Combining Equations 10.30 and 10.32, we can restate the completeness requirement purely in terms of the mental state:

$$p(w \mid m_t) = p(w \mid \boldsymbol{m}_{1:t}, \boldsymbol{c}_{1:t}). \tag{10.33}$$

This imposes a true constraint on the metalevel MDPs we can marginalize. Specifically, Equation 10.32 says that the mental state must be a *sufficient statistic* for the full history of mental states and computations with respect to the world state. To see why this property is necessary, note that Markov decision processes must satisfy the Markov property. That is, the probability of the next state must depend only on the current state and action; formally,

$$p(m_{t+1} \mid m_t, c_t) = p(m_{t+1} \mid \boldsymbol{m}_{1:t}, \boldsymbol{c}_{1:t}). \tag{10.34}$$

We can then show that this property implies Equation 10.33. Making explicit the marginalization over w on each side of the equation, we have

$$\sum_w p(m_{t+1} \mid m_t, c_t, w)p(w \mid m_t, c_t)$$

$$= \sum_w p(m_{t+1} \mid \boldsymbol{m}_{1:t}, \boldsymbol{c}_{1:t}, w)p(w \mid \boldsymbol{m}_{1:t}, \boldsymbol{c}_{1:t}). \tag{10.35}$$

Next, we note that $p(m_{t+1} \mid \boldsymbol{m}_{1:t}, \boldsymbol{c}_{1:t}, w) = p(m_{t+1} \mid m_t, c_t, w) = T(m_{t+1} \mid m_t, c_t, w)$ by Equation 10.7. That is, the full (non-marginalized) transition function satisfies the Markov property. This gives us

$$\sum_w T(m_{t+1} \mid m_t, c_t, w)p(w \mid m_t, c_t) = \sum_w T(m_{t+1} \mid m_t, c_t, w)p(w \mid \boldsymbol{m}_{1:t}, \boldsymbol{c}_{1:t}).$$

$$\tag{10.36}$$

And from this it is clear that

$$p(w \mid m_t, c_t) = p(w \mid \boldsymbol{m}_{1:t}, \boldsymbol{c}_{1:t}). \tag{10.37}$$

Finally, noting that $p(w \mid m_t, c_t) = p(w \mid m_t)$, we arrive at Equation 10.33. Thus, the Markov property (Equation 10.34) implies the sufficiency of the mental state (Equation 10.33), meaning that we cannot have the Markov property without ensuring that the mental state is a sufficient statistic. This in turn means that the marginalized metalevel MDP can be a *Markov* decision process only if the mental state is a sufficient statistic. Note, however, that if we allowed the metalevel transition function to depend on the policy, one could define a variant of metalevel MDPs that drop the sufficiency requirement (see Fox & Tishby, 2012).

10.8   Identifying good metalevel policies

Here, we discuss a few general methods for identifying optimal (or at least reasonable) policies for metalevel MDPs. Following Equation 10.5, the optimal metalevel policy can be expressed as

$$\pi^*(m) = \text{Uniform}\left(\arg\max_c Q^*(m, c)\right). \tag{10.38}$$

Each of the methods below provides a different way to compute or approximate Q^*.

HISTORICAL NOTE: THE VALUE OF COMPUTATION

Historically (e.g. Russell and Wefald, 1991b), rational metareasoning has been defined in terms of a slightly different quantity, the *value of computation* (VOC). The VOC is exactly what it sounds like; it specifies the value of performing a computation, where "value" refers to the long-term value in the same sense as the action value function, Q^*. However, the VOC specifically refers to the *increase* in reward one would gain by computing instead of deciding immediately. That is,

$$\text{VOC}(m, c) = Q^*(m, c) - R(m, \perp). \tag{10.39}$$

One advantage of this formulation is that we can define the optimal termination rule as executing $\perp$ whenever no computation has positive VOC. However, the Q function will be more familiar to most researchers, and is easier to work with in practice. For this reason, we will use the Q function throughout. Note, however, that the only difference between the two functions is constant with respect to c. Thus, maximizing either will yield the optimal policy (we can replace Q^* with VOC in Equation 10.38).

10.8.1 Backward induction

For metalevel MDPs with sufficiently small state spaces, the most robust and accurate method for identifying an optimal policy is backward induction, a form of dynamic programming. See Puterman (2014) for a general overview of this method. Here, we provide a brief introduction and a few practical suggestions for applying this approach to metalevel MDPs.

Backward induction is a method for computing the optimal value functions, Q^* and V^*, of an MDP. It is based on recursive definitions of the optimal value functions:

$$Q^*(s, a) = R(s, a) + \mathop{\mathbb{E}}_{s' \sim T(s,a)} \left[V^*(s') \right],$$
$$V^*(s) = \max_a Q^*(s, a).$$
(10.40)

These are referred to as *Bellman equations*. Backward induction is an especially simple and efficient application of the Bellman equations for MDPs that are finite and acyclic—that is, there are a finite number of states and one cannot visit the same state twice within a single episode. In such cases, we can assume (without loss of generality) that there is a single absorbing terminal state $s_\perp$ whose value is $V^*(s_\perp) = 0$, by definition. Because there are a finite number of states and no state can be reached from itself, any invocation of Q^* or V^* must eventually hit $V^*(s_\perp) = 0$, the base case.

In a metalevel MDP, it is more natural to define the base case with $Q^*(m, \perp) = R(m, \perp)$.[11] The value functions can then be defined:

$$Q^*(m, c) = \begin{cases} R(m, \perp) & \text{if } c = \perp \\ \mathbb{E}_{m' \sim T(m,c)} \left[V^*(m') \right] - \text{cost}(m, c) & \text{otherwise,} \end{cases}$$
(10.41)
$$V^*(m) = \max_a Q^*(m, c).$$

These equations can be directly implemented as a recursive program, as illustrated in Figure 10.4. Note, however, that a naive implementation may compute the value of single mental state many times if it can be reached in multiple ways. To prevent this, we *memoize* the value function. This means that if the function is called twice with the same argument, it will return the

---

11. This simply brings the base case up one level in the recursion, as executing $\perp$ always results in the terminal state.

```
function Q(m::MentalState, c::Computation)
    if c == TERM
        term_reward(m) # R(m, ⊥)
    else
        # E_{m' ∼ T(m,c)} [V*(m')] − cost(m, c)
        sum(p * V(m_next) for (p, m_next) in transition(m, c)) - cost(m, c)
    end
end

@memoize function V(m::MentalState)
    # max_{c∈C(m)} Q(m, c)
    maximum(Q(m, c) for c in computations(m))
end
```

FIGURE 10.4. Recursive implementation of backward induction in Julia.

result from the first call, rather than recomputing it. The combination of memoization and recursion is the defining feature of a "top-down" implementation of backward induction. It ensures that V^* is called on each mental state exactly once (assuming that all mental states are reachable from the initial mental state).

IMPLEMENTATION CONCERNS

Here, we address a few practical concerns that arise when applying backward induction to metalevel MDPs.

- Backward induction only applies for discrete state spaces. If the mental state space is continuous (and low-dimensional), one can discretize it. That is, we divide the space into evenly sized bins and create one state for the center of each bin. The discretized transition function is identical to the original transition function except that it "rounds" the generated mental state to the nearest discretized state. To compute an explicit probability mass function, one must integrate over each bin to compute the probability of transitioning to the corresponding state.
- The state space must be finite in addition to discrete. In many cases, however, the natural state space will be unbounded. To address this, we can impose a bound on the number of computations that can be performed (ideally, this bound will be also imposed on the experimental task one is modeling, e.g., by a time limit). The bound is implemented by adding a timestep to the mental state and removing all computations except ⊥ from the set of possible computations in mental states with maximal timestep. Note that adding the timestep counter also ensures that the MDP is acyclic.
- The state spaces of metalevel MDPs often have symmetry structure that reduces the effective size of the state space. For example, in choice

tasks, the order of the items is irrelevant. One way to implement this is to define a hash function that returns the same value for mental states that are functionally identical (specifically, that must have the same $V^*(s)$). This hash function can be used for memoization such that the value of a set of functionally identical states is only computed once (with the memoized value returned for all other states in the set). In Section 6.2.3, we use a hash function that processes a decision tree recursively, using a commutative operation (summation) to combine the keys for subtrees whose roots are siblings.

- Although the "top-down" implementation of backward induction (Figure 10.4) is simple, perhaps even beautiful, it is not the most efficient. When speed is a concern—and it almost always is—a "bottom-up" implementation is usually preferred. Such an implementation explicitly iterates over the state space, beginning with all states at the maximal time step and proceeding backward. See Puterman (2014) for further details on the algorithm.

While backward induction can identify an exact optimal policy (or approximate it to arbitrary precision), it is only tractable when the (discretized) mental state space is small enough that one can iterate over every state in a reasonable amount of time. When the state space is too large, one must turn to approximate solutions. We discuss several possibilities in the following sections.

10.8.2 The myopic policy

In their pioneering work on metareasoning Russell and Wefald (1991b) suggested an approximation to rational metareasoning by one-step lookahead, what they called the metalevel greedy approximation. This myopic (or "meta-greedy") policy can be defined as

$$\pi^{\text{myopic}}(m) = \text{Uniform}\left(\arg\max_c Q^{\text{myopic}}(m, c)\right), \qquad (10.42)$$

where

$$Q^{\text{myopic}}(m, c) = \mathop{\mathbb{E}}_{m' \sim T(m,c)}\left[R(m', \perp)\right] - \text{cost}(m, c), \qquad (10.43)$$

with $Q^{\text{myopic}}(m, \perp) = R(m', \perp)$. The myopic action value function Q^{myopic} gives the expected termination reward after performing one more computation, less the cost of that computation. Thus, the myopic policy selects each computation as if it will be the last one executed.

The myopic policy is often a decent approximation, and displays reasonable behavior in many cases. The problem with it is that it systematically underestimates the value of computation, which leads it to stop computing too early. We can see this easily in the tallying example. If one has the mental state $(3, 1)$, then considering one additional cue cannot change the decision one would make. The next mental state will be either $(4, 1)$ or $(3, 2)$, and car X will be in the lead in either case. Clearly there is no benefit to considering a single additional cue, so the myopic policy will terminate. However, given the opportunity to consider several more cues, the tide could easily be swayed in favor of the other car. If computation were not very costly, it would likely be worth computing more.

10.8.3   Multistep lookahead

In their analysis of economic information-seeking, Gabaix and Laibson (2005) took note of the myopic policy's struggle with premature termination. They proposed a possible solution in their *directed cognition* model. This model can be understood as a metalevel policy that selects *sequences* of computations rather than computations (exactly like options in hierarchical reinforcement learning; Sutton et al., 1999). Although a single computation may not be sufficient to change one's decision starting from the mental state $(3, 1)$, a sequence of three computations could yield $(3, 4)$, a mental state in which the agent would switch to selecting a_2. If the cost is sufficiently low, the expected value of executing that sequence will be higher than the expected value of terminating, and the directed cognition policy would continue to compute.

When applying this strategy, one need not (and generally should not) actually commit to taking the full sequence of computations that previously had maximal value. This is because the outcome of the first computation may make a different computation more valuable. Note, however, that it results in a dissociation between the agent's implicit assumptions when selecting computations (that it will execute all the computations in the sequence) and reality (that it can choose not to complete the sequence). More precisely, the agent is assuming that it will have less control over future computations than they actually will. This is precisely the same situation as the myopic policy was in (assuming that it would always terminate on the next step), but less severe.

Hay et al. (2012) proposed another solution to the early stopping problem, the *blinkered approximation*. Like directed cognition, the blinkered approximation engages in a sort of lookahead that reduces the flexibility in choosing future computations. Unlike directed cognition, however, the blinkered approximation takes into account the fact that it can choose to adjust its plan based on the outcome of each computation. Glossing over details, the value of a computation is approximated by its value in a smaller metalevel MDP that includes only the computations that reason about the expected utility of the same external action. In this way, the solution to a large metalevel MDP is approximated by the composition of solutions to many smaller metalevel MDPs. The blinkered approximation is always at least as good as (and often better than) the direction cognition and myopic policies. However it can only be applied in cases where a small number of computations are relevant to each action. And even then, it cannot account for the synergistic value of learning about multiple external actions.

All of the strategies we have discussed so far are *model-based*. That is, they estimate the value of computations by simulating their possible outcomes. This can be contrasted with *model-free* strategies that learn policies by trial and error. In the following section, we describe an algorithm that we developed, which combines model-based and model-free reasoning to quickly identify high-performing metalevel policies (Callaway et al., 2018).

10.9   Bayesian metalevel policy search

Bayesian metalevel policy search, or BMPS, is an algorithm that learns policies for metalevel MDPs. At its core, BMPS is a reinforcement learning (RL) algorithm. And in principle, any reinforcement-learning method could be applied to learn metalevel policies. However, metalevel MDPs pose an especially challenging type of problem for typical RL algorithms. In particular, metalevel MDPs present an extreme form of the *credit assignment problem*. In each episode, the agent takes many computations, but receives only a single external reward. If the agent receives a large termination reward, it is unclear which of the many executed computations were important for producing that good outcome and which could have been skipped. This makes it hard to learn which computations are worth performing. To make matters worse, the termination reward often depends greatly on factors outside of the agent's control. If one is choosing between many bad options, the agent cannot get a good termination reward, no matter how well they compute. Together, these factors make metalevel reinforcement learning very challenging.

The BMPS algorithm attempts to make the learning problem easier by endowing the agent with rich knowledge about the structure of the problem. To do so, it draws on work in rational metareasoning aiming to quantify and understand the value of computation (Matheson, 1968; Horvitz, 1987; Russell and Wefald, 1991b). Specially, BMPS draws on work quantifying the *value of information* generated by computation.

10.9.1   The value of information

We define the value of information (VOI) as the expected utility of a decision you could make based on that information. (Note that, like VOC, VOI is usually defined relative to the utility of acting without information; we drop the baseline here for the same reasons as before.) For information we already have, this is simply the termination reward. Thus, we can write

$$\text{VOI}(b) = R(b, \perp) = \max_{a} \; \mathbb{E}_{w \sim b} \left[U(w, a) \right], \qquad (10.44)$$

using the termination reward defined in Equation 10.27. Note that VOI can only be used when taking the Bayesian view of computation; therefore, we will not distinguish between mental states and beliefs, using b for both.

VOI is most useful for quantifying the value of information that *we don't have yet*. More precisely, VOI quantifies the expected value of a belief state we will have after gathering more information. Given a distribution of possible future belief states, B', the VOI is defined as

$$\text{VOI}(B') = \mathbb{E}_{b' \sim B'} \left[R(b', \perp) \right]. \qquad (10.45)$$

Using this notation, we can express the Q^* function as

$$Q^*(b_t, c_t) = \text{VOI}(B_N) - \mathbb{E} \left[\sum_{i=t}^{N} \text{cost}(b_i, c_i) \right], \qquad (10.46)$$

where B_N is the distribution of terminal beliefs (assuming that one begins by executing c_t in b_t and follows the optimal policy from then on). The challenge of course is that we do not know what this distribution is. However, we can put bounds on it.

10.9.2   Bounding the value of information

As illustrated in Figure 10.5, the value of information acquired by an optimal metalevel policy increases monotonically with the number of computations it performs. Moreover, it is bounded by two values, corresponding to the VOI associated with a distribution of possible beliefs.

The minimum VOI is the value of information produced by the first computation. In this case, the distribution of beliefs is simply $B' = T(b, c)$. Thus, the myopic value of information is defined as

$$\text{VOI}_{\text{myopic}}(b, c) = \text{VOI}(T(b, c)) = \mathbb{E}_{b' \sim T(b,c)} \left[R(b', \perp) \right].$$

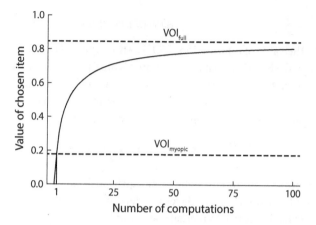

FIGURE 10.5. **Illustration of the value of information features.** The solid line shows the average value of the item chosen after different numbers of computations selected by a near-optimal policy for the metalevel MDP defined in Section 6.2.2, assuming no computational costs. The dashed lines show values for the two VOI features in the initial belief state: $\text{VOI}_{\text{myopic}}$ is the value after one computation and VOI_{full} is the asymptotic value after infinite computations.

The maximum VOI is the value of full, or "perfect" information (Howard, 1966). In this case, each of the possible future beliefs assigns all probability to a single world state. Denote such a belief as $b_w^* = \text{Uniform}(\{w\})$. The distribution over these beliefs, B_b^*, is defined as $\Pr(B_b^* = b_w^*) = b(w)$. Intuitively, the subjective probability of coming to the belief that assigns all probability to state w is equal to the subjective probability that w is in fact the true world state. We can then define the value of full information as

$$\text{VOI}_{\text{full}}(b) = \text{VOI}(B_b^*) = \mathop{\mathbb{E}}_{w \sim b} \left[R(b_w^*, \bot) \right]. \tag{10.47}$$

10.9.3 Learning to select computations

Given that $\text{VOI}_{\text{myopic}}(b, c)$ and $\text{VOI}_{\text{full}}(b)$ provide lower and upper bounds on the value of information, it follows that the true optimal value of information ($\text{VOI}(B_N)$ in Equation 10.46) is an interpolation between these two values. This suggests the following approximation,

$$Q^{\text{bmps}}(b, c; \boldsymbol{\beta}) = \beta_1 \text{VOI}_{\text{myopic}}(b, c) + \beta_2 \text{VOI}_{\text{full}}(b) - (\text{cost}(b, c) + \beta_{\text{cost}}),$$

$$\tag{10.48}$$

where β_1 and β_2 are the interpolation weights (with $\beta_1 + \beta_2 = 1$) and $\beta_{\text{cost}} \geq 0$ captures the expected cost of future computations. In specific domains, we may be able to identify additional VOI features that quantify the value of acquiring an intermediate amount of information. For example, in Section 6.2.2 we use a VOI feature for learning the exact value of a single item in the choice set. One can add as many features as one likes, adding an additional weight for each, and ensuring that they all sum to 1. Even with additional features, the approximation makes the very strong assumption that the interpolation weights and expected future cost are constant across belief states, an assumption that will almost certainly not hold. However, it turns out that this rough approximation is sufficient to produce near-optimal metalevel policies in some cases. This policy is defined as

$$\pi^{\text{bmps}}(s; \boldsymbol{\beta}) = \text{Uniform}\left(\arg\max_a Q^{\text{bmps}}(s, a; \boldsymbol{\beta})\right). \tag{10.49}$$

How should the weights be determined? Treating the approximation literally, the most appropriate strategy would be Q-learning, which tries to find weights that predict the actual returns of the policy. However, because the approximation is so rough, this turns out not to work well in practice. Instead, we can treat the weights as arbitrary parameters of a policy and use policy search to find weights that maximize performance. That is

$$\boldsymbol{\beta}^* = \arg\max_{\boldsymbol{\beta}} \mathbb{E}\left[\sum_{t=1}^{N} R(m_t, c_t, w) \,\middle|\, c_t \sim \pi^{\text{bmps}}(m_t; \boldsymbol{\beta}).\right] \tag{10.50}$$

The key advantage of BMPS over a more generic RL algorithm is that the policy has a very small number of parameters. This allows us to use an efficient search strategy such as Bayesian optimization. Callaway et al. (2018) present an evaluation of BMPS trained with Bayesian optimization on several benchmark tasks, finding that it outperforms the blinkered policy described above as well as a generic deep reinforcement learning algorithm (DQN). In a cognitive modeling application, consistency may be more important than efficiency; in these cases, we can use exhaustive search algorithms that produce similar results when run multiple times. See Callaway et al. (2021, pg. 21) for an example of such a method.

ACKNOWLEDGMENTS

The research surveyed in this book builds on and is inspired by the excellent work of many insightful researchers from many different disciplines. The citations throughout this book acknowledge some relevant contributions of some of them. There are likely many others not cited in this book. This is by no means a reflection of our assessment of their work's quality, relevance, or influence. It is merely a side effect of our time-constrained writing process, whose sole goal was to make our prior work accessible to a broader audience.

Several chapters of this book reuse some of our previous writing. Concretely, Chapter 4 and Chapter 6 reuse material from Lieder et al. (2024a) and Chapter 5 reuses material from Lieder (2018b).

We have been fortunate to have numerous collaborators who have helped us formulate the theory of resource rationality, develop and evaluate resource-rational models of human cognition, and devise and test interventions for enhancing human rationality. We are deeply grateful for their contributions to this work.

We are also grateful to the many colleagues who have provided critical feedback and constructive suggestions on the research surveyed in this book. In addition, we thank Ham Huang, Ruaridh Mon-Williams, Daniel Reichman, Cameron Turner, Bonan Zhao, Jian-Qiao Zhu, and two anonymous reviewers for their comments on the manuscript.

We thank the funders who have supported the research summarized in this book, including the National Science Foundation, the Templeton World Charity Foundation, the Office of Naval Research, the Air Force Office of Scientific Research, the University of California, the Max Planck Society, the Cyber Valley Research Fund, and the NOMIS Foundation.

We would also like to thank our editor Hallie Stebbins for her enthusiasm for the project, Chloe Coy for shepherding us through the publication process, and Bhisham Bherwani for his careful proofreading and copyediting. We are also grateful to the anonymous reviewers for their constructive feedback and thoughtful comments. Finally, we would like to thank Princeton University Press for publishing this book, giving us the opportunity to write it, allowing us to make PDFs of our drafts of the books' chapters freely available on the internet, and supporting us along the way.

BIBLIOGRAPHY

Abbott, J. T. and Griffiths, T. L. (2011). Exploring the influence of particle filter parameters on order effects in causal learning. *Proceedings of the 33rd Annual Conference of the Cognitive Science Society.*

Acharya, A., Chen, X., Lewis, R. L., and Howes, A. (2017). Human visual search as a deep reinforcement learning solution to a POMDP. Proceedings of the 39th Annual Meeting of the Cognitive Science Society.

Agrawal, M., Mattar, M. G., Cohen, J. D., and Daw, N. D. (2022). The temporal dynamics of opportunity costs: A normative account of cognitive fatigue and boredom. *Psychological Review,* 129(3):564–585.

Allais, M. (1953). Le comportement de l'homme rationnel devant le risque: Critique des postulats et axiomes de l'école Américaine. *Econometrica,* 21(4):503–546.

Allcott, H. and Kessler, J. B. (2019). The welfare effects of nudges: A case study of energy use social comparisons. *American Economic Journal: Applied Economics,* 11(1):236–276.

Amit, R., Meir, R., and Ciosek, K. (2020). Discount factor as a regularizer in reinforcement learning. In *Proceedings of the 37th International Conference on Machine Learning.*

Anderson, J. R. (1978). Arguments concerning representations for mental imagery. *Psychological Review,* 85(4):249–249.

Anderson, J. R. (1990). *The Adaptive Character of Thought.* Erlbaum.

Anderson, J. R. (1991). The adaptive nature of human categorization. *Psychological Review,* 98(3):409–429.

Anderson, J. R. (1996). ACT: A simple theory of complex cognition. *American Psychologist,* 51(4):355–365.

Anderson, J. R. (2013). *The Architecture of Cognition.* Psychology Press.

Anderson, J. R. and Milson, R. (1989). Human memory: An adaptive perspective. *Psychological Review,* 96(4):703–719.

Anderson, J. R. and Schooler, L. J. (1991). Reflections of the environment in memory. *Psychological Science,* 2(6):396–408.

Anderson, J. R. and Sheu, C.-F. (1995). Causal inferences as perceptual judgments. *Memory & Cognition,* 23:510–524.

Ariely, D. (2009). *Predictably Irrational.* HarperCollins.

Armel, K. C., Beaumel, A., and Rangel, A. (2008). Biasing simple choices by manipulating relative visual attention. *Judgment and Decision Making,* 3(5):396–403.

Augier, M. and March, J. G. (2004). *Models of a Man: Essays in Memory of Herbert A. Simon.* MIT Press.

Bahrini, A., Khamoshifar, M., Abbasimehr, H., Riggs, R. J., Esmaeili, M., Majdabadkohne, R. M., and Pasehvar, M. (2023). ChatGPT: Applications, opportunities, and threats. 2023 Systems and Information Engineering Design Symposium.

Banino, A., Balaguer, J., and Blundell, C. (2021). PonderNet: Learning to Ponder. arXiv preprint *arXiv:2107.05407*.

Bao, J. and Ho, B. (2015). Heterogeneous effects of informational nudges on pro-social behavior. *The BE Journal of Economic Analysis & Policy*, 15(4):1619–1655.

Bari, B. A. and Gershman, S. J. (2024). Resource-rational psychopathology. Behavioral Neuroscience, 138(4), 221–234.

Baron, J. (1985). *Rationality and Intelligence*. Cambridge University Press.

Baron, J. (1997). Biases in the quantitative measurement of values for public decisions. *Psychological Bulletin*, 122(1):72–88.

Baron, J. (1998). *Judgment Misguided: Intuition and Error in Public Decision Making*. Oxford University Press.

Beach, L. R. and Mitchell, T. R. (1978). A contingency model for the selection of decision strategies. *Academy of Management Review*, 3(3):439–449.

Becker, F., Skirzyński, J., van Opheusden, B., and Lieder, F. (2022). Boosting human decision-making with AI-generated decision aids. *Computational Brain & Behavior*, 5(4): 467–490.

Becker, F., Wirzberger, M., Pammer-Schindler, V., Srinivas, S., and Lieder, F. (2023). Systematic metacognitive reflection helps people discover far-sighted decision strategies: A process-tracing experiment. *Judgment and Decision Making*, 18:e15.

Bekkers, R. and Wiepking, P. (2011). A literature review of empirical studies of philanthropy: Eight mechanisms that drive charitable giving. *Nonprofit and Voluntary Sector Quarterly*, 40(5):924–973.

Bentham, J. (1961). *An Introduction to the Principles of Morals and Legislation*. Doubleday, Garden City. Originally published in 1789.

Bernoulli, D. (1738/1954). Exposition of a new theory on the measurement of risk. *Econometrica*, 22(1):22–36.

Bhui, R. and Gershman, S. J. (2018). Decision by sampling implements efficient coding of psychoeconomic functions. *Psychological Review*, 125(6):985–1001.

Bhui, R., Lai, L., and Gershman, S. J. (2021). Resource-rational decision making. *Current Opinion in Behavioral Sciences*, 41:15–21.

Binz, M. and Schulz, E. (2022). Modeling human exploration through resource-rational reinforcement learning. *Advances in Neural Information Processing Systems* 35:31755–31768.

Binz, M. and Schulz, E. (2023). Reconstructing the Einstellung effect. *Computational Brain & Behavior*, 6(3):526–542.

Blumenthal-Barby, J. S. and Krieger, H. (2015). Cognitive biases and heuristics in medical decision making: A critical review using a systematic search strategy. *Medical Decision Making*, 35(4):539–557.

Bolton, P. and Dewatripont, M. (2004). *Contract Theory*. MIT press.

Bonawitz, E., Denison, S., Gopnik, A., and Griffiths, T. (2014a). Win-stay, lose-sample: A simple sequential algorithm for approximating Bayesian inference. *Cognitive Psychology*, 74:35–65.

Bonawitz, E., Denison, S., Griffiths, T. L., and Gopnik, A. (2014b). Probabilistic models, learning algorithms, and response variability: Sampling in cognitive development. *Trends in Cognitive Sciences*, 18(10):497–500.

Börgers, T. (2015). *An Introduction to the Theory of Mechanism Design*. Oxford University Press.

Bornstein, B. H. and Emler, A. C. (2001). Rationality in medical decision making: A review of the literature on doctors' decision-making biases. *Journal of Evaluation in Clinical Practice*, 7(2):97–107.

Botvinick, M. M. and Cohen, J. D. (2014). The computational and neural basis of cognitive control: Charted territory and new frontiers. *Cognitive Science*, 38(6):1249–1285.

Botvinick, M. M., Niv, Y., and Barto, A. C. (2009). Hierarchically organized behavior and its neural foundations: A reinforcement learning perspective. *Cognition*, 113(3): 262–280.

Boureau, Y.-L., Sokol-Hessner, P., and Daw, N. D. (2015). Deciding how to decide: Self-control and meta-decision making. *Trends in Cognitive Sciences*, 19(11):700–710.

Bovens, L. (2009). The Ethics of Nudge. In Grüne-Yanoff, T. and Hansson, S.O. (Eds.) Preference Change. Theory and Decision Library, vol 42. Springer, Dordrecht.

Bowers, J. and Davis, C. (2012). Bayesian just-so stories in psychology and neuroscience. *Psychonomic Bulletin and Review*, 138(3):389–414.

Bramley, N. R., Dayan, P., Griffiths, T. L., and Lagnado, D. A. (2017). Formalizing Neurath's ship: Approximate algorithms for online causal learning. *Psychological Review*, 124(3):301.

Braver, T. (2012). The variable nature of cognitive control: A dual mechanisms framework. *Trends in Cognitive Sciences*, 16(2):106–113.

Bröder, A. (2003). Decision making with the "adaptive toolbox": Influence of environmental structure, intelligence, and working memory load. *Journal of Experimental Psychology: Learning, Memory, and Cognition*, 29(4):611–625.

Bucher, T., Collins, C., Rollo, M. E., McCaffrey, T. A., De Vlieger, N., Van der Bend, D., Truby, H., and Perez-Cueto, F. J. (2016). Nudging consumers towards healthier choices: A systematic review of positional influences on food choice. *British Journal of Nutrition*, 115(12):2252–2263.

Buesing, L., Bill, J., Nessler, B., and Maass, W. (2011). Neural dynamics as sampling: A model for stochastic computation in recurrent networks of spiking neurons. *PLoS Computational Biology*, 7(11):e1002211.

Burga, T., Gross, P., Pons, E., Maier, M., Cheung, V., and Lieder, F. (2023). Decision-makers systematically overlook crucial considerations in social dilemmas: Implications for finding public policies that maximize collective well-being. Technical report.

Bustamante, L., Lieder, F., Musslick, S., Shenhav, A., and Cohen, J. (2021). Learning to overexert cognitive control in a Stroop task. *Cognitive, Affective, & Behavioral Neuroscience*, 21:453–471.

Butko, N. J. and Movellan, J. R. (2008). I-POMDP: An Infomax model of eye movement. IEEE International Conference on Development and Learning.

Bénabou, R. and Tirole, J. (2006). Incentives and prosocial behavior. *American Economic Review*, 96(5):1652–1678.

Callaway, F., Griffiths, T. L., Norman, K. A., and Zhang, Q. (2024). Optimal metacognitive control of memory recall. *Psychological Review*, 131(3):781–811.

Callaway, F., Gul, S., Krueger, P. M., Griffiths, T. L., and Lieder, F. (2018). Learning to select computations. *Proceedings of the 35th Conference on Uncertainty in Artificial Intelligence.*

Callaway, F., Hardy, M., and Griffiths, T. L. (2023). Optimal nudging for cognitively bounded agents: A framework for modeling, predicting, and controlling the effects of choice architectures. *Psychological Review,* 130(6):1457–1491.

Callaway, F., Jain, Y. R., van Opheusden, B., Das, P., Iwama, G., Gul, S., Krueger, P. M., Becker, F., Griffiths, T. L., and Lieder, F. (2022a). Leveraging artificial intelligence to improve people's planning strategies. *Proceedings of the National Academy of Sciences,* 119(12):e2117432119.

Callaway, F., Rangel, A., and Griffiths, T. L. (2021). Fixation patterns in simple choice reflect optimal information sampling. *PLoS Computational Biology,* 17(3):e1008863.

Callaway, F., van Opheusden, B., Gul, S., Das, P., Krueger, P. M., Lieder, F., and Griffiths, T. L. (2022b). Rational use of cognitive resources in human planning. *Nature Human Behaviour,* 6(8):1–14.

Camerer, C. F. and Fehr, E. (2006). When does "economic man" dominate social behavior? *Science,* 311(5757):47–52.

Camerer, C. F. and Hogarth, R. M. (1999). The effects of financial incentives in experiments: A review and capital-labor-production framework. *Journal of Risk and Uncertainty,* 19:7–42.

Caplin, A. and Dean, M. (2015). Revealed preference, rational inattention, and costly information acquisition. *American Economic Review,* 105(7):2183–2203.

Caplin, A., Dean, M., and Martin, D. (2011). Search and satisficing. *American Economic Review,* 101(7):2899–2922.

Capraro, V., Jagfeld, G., Klein, R., Mul, M., and de Pol, I. v. (2019). Increasing altruistic and cooperative behaviour with simple moral nudges. *Scientific Reports,* 9(1):11880.

Carlsson, F. and Johansson-Stenman, O. (2019). Optimal prosocial nudging. Technical report.

Caviola, L. and Greene, J. D. (2023). Boosting the impact of charitable giving with donation bundling and micromatching. *Science Advances,* 9(3):eade7987.

Caviola, L., Schubert, S., and Greene, J. D. (2021). The psychology of (in) effective altruism. *Trends in Cognitive Sciences,* 25(7):596–607.

Caviola, L., Schubert, S., and Nemirow, J. (2020). The many obstacles to effective giving. *Judgment and Decision Making,* 15(2):159–172.

Chan, S. W., Lu, S., Cheng, N. H. C., Chui, C. H.-K., and Yat-Sang Lum, T. (2024). Promoting charitable donations and volunteering through nudge tools from the perspective of behavioral economics: A systematic review. *Nonprofit and Voluntary Sector Quarterly,* 54(2):435–461.

Charnov, E. L. (1976). Optimal foraging, the marginal value theorem. *Theoretical Population Biology,* 9(2):129–136.

Chase, W. G. (1978). Elementary information processes. In W. K. Estes (Ed.), Handbook of Learning & Cognitive Processes: V. Human Information. Pages 19–90. Erlbaum.

Chater, N. (2015). Can cognitive science create a cognitive economics? *Cognition,* 135:52–55.

Chater, N. and Loewenstein, G. (2023). The I-frame and the S-frame: How focusing on individual-level solutions has led behavioral public policy astray. *Behavioral and Brain Sciences,* 46:e147.

Chen, H., Chang, H. J., and Howes, A. (2021). Apparently irrational choice as optimal sequential decision making. *Proceedings of the AAAI Conference on Artificial Intelligence,* 35(1):792–800.

Cheyette, S. J. and Piantadosi, S. T. (2020). A unified account of numerosity perception. *Nature Human Behaviour*, 4(12):1265–1272.

Choi, J. J., Laibson, D., Madrian, B. C., and Metrick, A. (2004). For better or for worse: Default effects and 401(k) savings behavior. In Perspectives on the Economics of Aging. Pages 81–126. University of Chicago Press.

Christensen-Szalanski, J. J. (1978). Problem solving strategies: A selection mechanism, some implications, and some data. *Organizational Behavior and Human Performance*, 22(2):307–323.

Christian, B. and Griffiths, T. L. (2016). *Algorithms to Live By: The Computer Science of Human Decisions*. Henry Holt and Company.

Collins, A. G. and Frank, M. J. (2013). Cognitive control over learning: Creating, clustering, and generalizing task-set structure. *Psychological Review*, 120(1):190–229.

Consul, S., Heindrich, L., Stojcheski, J., and Lieder, F. (2022a). Improving human decision-making by discovering efficient strategies for hierarchical planning. *Computational Brain & Behavior*, 5(2):185–216.

Consul, S., Stojcheski, J., and Lieder, F. (2022b). Leveraging AI for effective to-do list gamification. Gesellschaft für Informatik eV.

Correa, C. G., Ho, M. K., Callaway, F., Daw, N. D., and Griffiths, T. L. (2023). Humans decompose tasks by trading off utility and computational cost. *PLOS Computational Biology*, 19(6):e1011087.

Costermans, J., Lories, G., and Ansay, C. (1992). Confidence level and feeling of knowing in question answering: The weight of inferential processes. *Journal of Experimental Psychology: Learning, Memory, and Cognition*, 18(1):142–150.

Cox, R. (1946). Probability, frequency, and reasonable expectation. *American Journal of Physics*, 14:1–13.

Cox, R. T. (1961). *The Algebra of Probable Inference*. Johns Hopkins University Press.

Curşeu, P. L., Jansen, R. J., and Chappin, M. M. (2013). Decision rules and group rationality: Cognitive gain or standstill? *PloS One*, 8(2):e56454.

d'Abate, C. P. and Eddy, E. R. (2007). Engaging in personal business on the job: Extending the presenteeism construct. *Human Resource Development Quarterly*, 18(3):361–383.

Dasgupta, I. and Griffiths, T. L. (2022). Clustering and the efficient use of cognitive resources. *Journal of Mathematical Psychology*, 109:102675.

Dasgupta, I., Schulz, E., and Gershman, S. J. (2017). Where do hypotheses come from? *Cognitive Psychology*, 96:1–25.

Dasgupta, I., Schulz, E., Tenenbaum, J. B., and Gershman, S. J. (2020). A theory of learning to infer. *Psychological Review*, 127(3):412–441.

Daw, N. D., Niv, Y., and Dayan, P. (2005). Uncertainty-based competition between prefrontal and dorsolateral striatal systems for behavioral control. *Nature Neuroscience*, 8(12):1704–1711.

Dawes, R. M. (1980). Social dilemmas. Annual Review of Psychology, 31:169–193.

Dayan, P. and Daw, N. D. (2008). Decision theory, reinforcement learning, and the brain. *Cognitive, Affective, & Behavioral Neuroscience*, 8(4):429–453.

de Finetti, B. (1937/1964). Foresight: Its logical laws, its subjective sources. In *Studies in Subjective Probability*. Robert E. Krieger Publishing Co.

de Groot, A. D. (1965). *Thought and Choice in Chess*. De Gruyter Mouton.

de Montmort, P. R. (1713). *Essay d'analyse sur les jeux de hazard.* J. Quillau.

Denison, S., Bonawitz, E., Gopnik, A., and Griffiths, T. (2013). Rational variability in children's causal inferences: The sampling hypothesis. *Cognition,* 126(2):285–300.

Diamond, A. (2013). Executive functions. *Annual Review of Psychology,* 64:135–168.

Dickhaut, J., Rustichini, A., and Smith, V. (2009). A neuroeconomic theory of the decision process. *Proceedings of the National Academy of Sciences,* 106(52):22145–22150.

Dingemanse, M. (2020). Resource-rationality beyond individual minds: The case of interactive language use. *Behavioral and Brain Sciences,* 43:23–24.

Dolan, R. J. and Dayan, P. (2013). Goals and habits in the brain. *Neuron,* 80(2):312–325.

Donders, F. C. (1969). On the speed of mental processes. *Acta Psychologica,* 30:412–431.

Dragan, A. D., Lee, K. C., and Srinivasa, S. S. (2013). Legibility and predictability of robot motion. *Proceedings of the 8th ACM/IEEE International Conference on Human-Robot Interaction.*

Drugowitsch, J., Moreno-Bote, R., Churchland, A. K., Shadlen, M. N., and Pouget, A. (2012). The Cost of Accumulating Evidence in Perceptual Decision Making. *Journal of Neuroscience,* 32(11):3612–3628.

Dunlosky, J. and Lipko, A. R. (2007). Metacomprehension: A brief history and how to improve its accuracy. *Current Directions in Psychological Science,* 16(4):228–232.

Eakin, D. K. (2005). Illusions of knowing: Metamemory and memory under conditions of retroactive interference. *Journal of Memory and Language,* 52(4):526–534.

Ebeling, F. and Lotz, S. (2015). Domestic uptake of green energy promoted by opt-out tariffs. *Nature Climate Change,* 5(9):868–871.

Edwards, W. (1954). The theory of decision making. *Psychological Bulletin,* 51(4):380–417.

Enke, B. (2024). The cognitive turn in behavioral economics. Technical report.

Erev, I. and Barron, G. (2005). On adaptation, maximization, and reinforcement learning among cognitive strategies. *Psychological Review,* 112(4):912–931.

Evans, J. S. B. T. (2003). In two minds: Dual-process accounts of reasoning. *Trends in Cognitive Sciences,* 7(10):454–459.

Evans, J. S. B. (2008). Dual-processing accounts of reasoning, judgment, and social cognition. *Annual Review of Psychology,* 59:255–278.

Evans, J. S. B. T. and Stanovich, K. E. (2013). Dual-process theories of higher cognition: Advancing the debate. *Perspectives on Psychological Science,* 8(3):223–241.

Farhi, E. and Gabaix, X. (2020). Optimal taxation with behavioral agents. *American Economic Review,* 110(1):298–336.

Fehr, E. and Falk, A. (2002). Psychological foundations of incentives. *European Economic Review,* 46(4–5):687–724.

Fiedler, K. (1988). The dependence of the conjunction fallacy on subtle linguistic factors. *Psychological Research,* 50(2):123–129.

Fiser, J., Berkes, P., Orbán, G., and Lengyel, M. (2010). Statistically optimal perception and learning: from behavior to neural representations. *Trends in Cognitive Sciences,* 14(3):119–130.

Fox, R. and Tishby, N. (2012) Bounded Planning in Passive POMDPs. *Proceedings of the 29th International Conference on Machine Learning.*

Fränken, J.-P., Theodoropoulos, N. C., and Bramley, N. R. (2022). Algorithms of adaptation in inductive inference. *Cognitive Psychology*, 137:101506.

Fum, D. and Del Missier, F. (2001). Adaptive selection of problem solving strategies. Proceedings of the 23rd Annual Conference of the Cognitive Science Society.

Gabaix, X. (2019). Behavioral inattention. In Handbook of Behavioral Economics: Applications and Foundations 1. Volume 2, pages 261–343. Elsevier.

Gabaix, X. (2020). A behavioral new Keynesian model. *American Economic Review*, 110(8):2271–2327.

Gabaix, X. (2023). Marshall lecture 2023: Behavioral macroeconomics via sparse dynamic programming. *Journal of the European Economic Association*, 21(6):2327–2376.

Gabaix, X. and Laibson, D. (2005). Bounded rationality and directed cognition. Technical report.

Gabaix, X. and Laibson, D. (2017). Myopia and discounting. Technical report.

Gailliot, M. T., Baumeister, R. F., DeWall, C. N., Maner, J. K., Plant, E. A., Tice, D. M., Brewer, L. E., and Schmeichel, B. J. (2007). Self-control relies on glucose as a limited energy source: willpower is more than a metaphor. *Journal of Personality and Social Psychology*, 92(2):325.

Gainsburg, I., Pauer, S., Abboub, N., Aloyo, E. T., Mourrat, J.-C., and Cristia, A. (2023). How effective altruism can help psychologists maximize their impact. *Perspectives on Psychological Science*, 18(1):239–253.

Geman, S., Bienenstock, E., and Doursat, R. (1992). Neural networks and the bias-variance dilemma. *Neural Computation*, 4:1–58.

Gershman, S. and Bhui, R. (2020). Rationally inattentive intertemporal choice. *Nature Communications*, 11(1):3365.

Gershman, S. J., Horvitz, E. J., and Tenenbaum, J. B. (2015). Computational rationality: A converging paradigm for intelligence in brains, minds, and machines. *Science*, 349(6245):273–278.

Gershman, S. J., Vul, E., and Tenenbaum, J. B. (2012). Multistability and perceptual inference. *Neural Computation*, 24(1):1–24.

Geweke, J. (1989). Bayesian inference in econometric models using Monte Carlo integration. *Econometrica*, 57(6):1317–1339.

Ghesla, C., Grieder, M., and Schmitz, J. (2019). Nudge for good? Choice defaults and spillover effects. *Frontiers in Psychology*, 10:178.

Gigerenzer, G. (1991). How to make cognitive illusions disappear: Beyond "heuristics and biases." *European Review of Social Psychology*, 2(1):83–115.

Gigerenzer, G. (2000). *Adaptive thinking: Rationality in the Real World*. Oxford University Press.

Gigerenzer, G. and Brighton, H. (2009). Homo heuristicus: Why biased minds make better inferences. *Topics in Cognitive Science*, 1(1):107–143.

Gigerenzer, G. and Gaissmaier, W. (2011). Heuristic decision making. *Annual Review of Psychology*, 62(1):451–482.

Gigerenzer, G. and Goldstein, D. G. (1996). Reasoning the fast and frugal way: Models of bounded rationality. *Psychological Review*, 103(4):650–669.

Gigerenzer, G. and Goldstein, D. G. (1999). Betting on one good reason: The take the best heuristic. In Gigerenzer, G. and Todd, P. (Eds.) *Simple Heuristics That Make Us Smart*. Oxford University Press.

Gigerenzer, G. and Hoffrage, U. (1995). How to improve Bayesian reasoning without instruction: Frequency formats. *Psychological Review*, 102(4):684–704.

Gigerenzer, G. and Selten, R. (2002). *Bounded Rationality: The Adaptive Toolbox*. MIT Press.

Gigerenzer, G. and Todd, P. M. (1999a). Fast and frugal heuristics: The adaptive toolbox. In
Simple Heuristics That Make Us Smart. Pages 3–34. Oxford University Press.

Gigerenzer, G. and Todd, P. M. (1999b). *Simple Heuristics That Make Us Smart*. Oxford
University Press, USA.

Gilks, W. R., Richardson, S., and Spiegelhalter, D. J. (1996). *Markov Chain Monte Carlo in
Practice*. Chapman & Hall.

Gilovich, T., Griffin, D., and Kahneman, D. (2002). *Heuristics and Biases: The Psychology of
Intuitive Judgment*. Cambridge University Press.

Girotra, K., Terwiesch, C., and Ulrich, K. T. (2010). Idea generation and the quality of the best
idea. *Management Science*, 56(4):591–605.

Gläscher, J., Daw, N., Dayan, P., and O'Doherty, J. (2010). States versus rewards: Dissocia-
ble neural prediction error signals underlying model-based and model-free reinforcement
learning. *Neuron*, 66(4):585–595.

Glimcher, P. W. and Fehr, E. (2013). *Neuroeconomics: Decision Making and the Brain*. Academic
Press, 2nd edition.

Godfrey-Smith, P. (2001). Three kinds of adaptationism. In S. H. Orzack and E. Sober (Eds.),
Adaptationism and Optimality. Pages 335–357. Cambridge University Press.

Gong, T. and Bramley, N. R. (2023). Continuous time causal structure induction with preven-
tion and generation. *Cognition*, 240:105530.

Gong, T., Gerstenberg, T., Mayrhofer, R., and Bramley, N. R. (2023). Active causal structure
learning in continuous time. *Cognitive Psychology*, 140:101542.

Good, I. J. (1952). Rational Decisions. *Journal of the Royal Statistical Society: Series B (Methodological)*, 14(1):107–114.

Goodwin, T. (2012). Why we should reject "nudge." *Politics*, 32(2):85–92.

Gottlieb, J., Oudeyer, P. Y., Lopes, M., and Baranes, A. (2013). Information-seeking, curiosity,
and attention: Computational and neural mechanisms. *Trends in Cognitive Sciences*, 17(11),
585–593.

Grahek, I., Musslick, S., and Shenhav, A. (2020). A computational perspective on the roles of
affect in cognitive control. *International Journal of Psychophysiology*, 151:25–34.

Graves, A. (2016). Adaptive computation time for recurrent neural networks. arXiv preprint
arXiv:1603.08983.

Griffiths, T. L. (2020). Understanding human intelligence through human limitations. *Trends
in Cognitive Sciences*, 24(11):873–883.

Griffiths, T. L., Chater, N., Kemp, C., Perfors, A., and Tenenbaum, J. (2010). Probabilistic models of cognition: Exploring the laws of thought. *Trends in Cognitive Sciences*,
14:357–364.

Griffiths, T. L., Chater, N., and Tenenbaum, J. B. (2024). *Bayesian Models of Cognition: Reverse
Engineering the Mind*. MIT Press.

Griffiths, T. L., Kemp, C., and Tenenbaum, J. B. (2008). Bayesian models of cognition. In
Ron Sun (Ed.), *The Cambridge Handbook of Computational Cognitive Modeling*. Cambridge
University Press.

Griffiths, T. L., Lieder, F., and Goodman, N. D. (2015). Rational use of cognitive resources: Levels of analysis between the computational and the algorithmic. *Topics in Cognitive Science*, 7(2):217–229.

Griffiths, T. L. and Tenenbaum, J. B. (2001). Randomness and coincidences: Reconciling intuition and probability theory. Pages 370–375. Erlbaum.

Griffiths, T. L. and Tenenbaum, J. B. (2005). Structure and strength in causal induction. *Cognitive Psychology*, 51(4):334–384.

Griffiths, T. L. and Tenenbaum, J. B. (2006). Optimal predictions in everyday cognition. *Psychological Science*, 17(9):767–773.

Griffiths, T. L. and Tenenbaum, J. (2011). Predicting the future as Bayesian inference: People combine prior knowledge with observations when estimating duration and extent. *Journal of Experimental Psychology: General*, 140(4):725–743.

Griffiths, T. L., Vul, E., and Sanborn, A. (2012). Bridging levels of analysis for probabilistic models of cognition. *Current Directions in Psychological Science*, 21(4):263–268.

Gruneberg, M. M., Monks, J., and Sykes, R. N. (1977). Some methodological problems with feeling of knowing studies. *Acta Psychologica*, 41(5):365–371.

Gureckis, T. M. and Markant, D. B. (2012). Self-directed learning: A cognitive and computational perspective. *Perspectives on Psychological Science*, 7(5):464–481.

Hafenbrädl, S., Waeger, D., Marewski, J. N., and Gigerenzer, G. (2016). Applied decision making with fast-and-frugal heuristics. *Journal of Applied Research in Memory and Cognition*, 5(2):215–231.

Hahn, M., Futrell, R., Levy, R., and Gibson, E. (2022). A resource-rational model of human processing of recursive linguistic structure. *Proceedings of the National Academy of Sciences*, 119(43):e2122602119.

Hammersley, D. C. and Handscomb, J. M. (1964). *Monte Carlo Methods*. Methuen.

Hamrick, J. B., Smith, K. A., Griffiths, T. L., and Vul, E. (2015). Think again? The amount of mental simulation tracks uncertainty in the outcome. Volume 1. Cognitive Science Society.

Hart, J. T. (1965). Memory and the feeling-of-knowing experience. *Journal of Educational Psychology*, 56(4):208–216.

Hastie, T., Tibshirani, R., and Friedman, J. (2009). *The Elements of Statistical Learning*. Springer.

Hastings, W. K. (1970). Monte Carlo sampling methods using Markov chains and their applications. *Biometrika*, 57(1):97–109.

Hay, N. (2016). Principles of metalevel control. PhD thesis.

Hay, N., Russell, S., Tolpin, D., and Shimony, S. (2012). Selecting computations: Theory and applications. Volume abs/1408.2048, pages 346–355. AUAI Press.

Hazy, T. E., Frank, M. J., and O'Reilly, R. C. (2006). Banishing the homunculus: Making working memory work. *Neuroscience*, 139(1):105–118.

He, R., Jain, Y. R., and Lieder, F. (2021). Measuring and modelling how people learn how to plan and how people adapt their planning strategies to the structure of the environment. International Conference on Cognitive Modeling.

He, R., Jain, Y. R., and Lieder, F. (2022). Have I done enough planning or should I plan more? NeurIPS Workshop on Metacognition in the Age of AI.

He, R. and Lieder, F. (2023). What are the mechanisms underlying metacognitive learning in the context of planning? *Proceedings of the 45th Annual Conference of the Cognitive Science Society.*

Heindrich, L., Consul, S., and Lieder, F. (2025). An intelligent tutor for planning in large partially observable environments. *International Journal of Artificial Intelligence in Education.*

Heindrich, L. and Lieder, F. (2025). Discovering the curriculum with AI: A proof-of-concept demonstration with an intelligent tutoring system for teaching project selection. https://arxiv.org/abs/2406.04082.

Herrnstein, R. J. and Loveland, D. H. (1975). Maximizing and matching on concurrent ratio schedules. *Journal of Experimental Analysis of Behavior*, 24(1):107–116.

Hershfield, H. E., Goldstein, D. G., Sharpe, W. F., Fox, J., Yeykelis, L., Carstensen, L. L., and Bailenson, J. N. (2011). Increasing saving behavior through age-progressed renderings of the future self. *Journal of Marketing Research*, 48(SPL):S23–S37.

Hertwig, R. and Grüne-Yanoff, T. (2017). Nudging and boosting: Steering or empowering good decisions. *Perspectives on Psychological Science*, 12(6):973–986.

Hertwig, R. and Grüne-Yanoff, T. (2021). Boosting and nudging: Two paths toward better financial decisions. *Financial Education and Risk Literacy.*

Ho, M. K., Abel, D., Correa, C. G., Littman, M. L., Cohen, J. D., and Griffiths, T. L. (2022). People construct simplified mental representations to plan. *Nature*, 606(7912):129–136.

Ho, M. K., Cohen, J. D., and Griffiths, T. L. (2023). Rational simplification and rigidity in human planning. *Psychological Science*, 34(11):1281–1292.

Ho, M. K. and Griffiths, T. L. (2022). Cognitive science as a source of forward and inverse models of human decisions for robotics and control. *Annual Review of Control, Robotics, and Autonomous Systems*, 5(1):33–53.

Hoffrage, U. and Gigerenzer, G. (1998). Using natural frequencies to improve diagnostic inferences. *Academic Medicine*, 73(5):538–540.

Hoffrage, U., Gigerenzer, G., Krauss, S., and Martignon, L. (2002). Representation facilitates reasoning: What natural frequencies are and what they are not. *Cognition*, 84(3):343–352.

Hoffrage, U., Krauss, S., Martignon, L., and Gigerenzer, G. (2015). Natural frequencies improve Bayesian reasoning in simple and complex inference tasks. *Frontiers in Psychology*, 6:1473.

Holmes, P. and Cohen, J. D. (2014). Optimality and some of its discontents: Successes and shortcomings of existing models for binary decisions. *Topics in Cognitive Science*, 6(2):258–278.

Hoppe, D. and Rothkopf, C. A. (2019). Multi-step planning of eye movements in visual search. *Scientific Reports*, 9(1):144.

Horvitz, E. J. (1987). Reasoning about beliefs and actions under computational resource constraints. *Proceedings of the Conference on Uncertainty in Artificial Intelligence*, Pages 301–324. AUAI Press.

Horvitz, E. J., Suermondt, H., and Cooper, G. (1989). Bounded conditioning: Flexible inference for decisions under scarce resources. *Proceedings of the Conference on Uncertainty in Artificial Intelligence*, pages 182–193. Elsevier.

Houston, A. I. and McNamara, J. M. (1999). *Models of Adaptive Behaviour: An Approach Based on State.* Cambridge University Press.

Howard, R. A. (1966). Information value theory. *IEEE Transactions on Systems Science and Cybernetics*, 2(1):22–26.

Howes, A., Lewis, R. L., and Vera, A. (2009). Rational adaptation under task and processing constraints: Implications for testing theories of cognition and action. *Psychological Review*, 116(4):717–751.

Howes, A., Warren, P. A., Farmer, G., El-Deredy, W., and Lewis, R. L. (2016). Why contextual preference reversals maximize expected value. *Psychological Review*, 123(4):368–391.

Huys, Q. J. M., Eshel, N., O'Nions, E., Sheridan, L., Dayan, P., and Roiser, J. P. (2012). Bonsai trees in your head: How the Pavlovian system sculpts goal-directed choices by pruning decision trees. *PLoS Computational Biology*, 8(3):e1002410.

Huys, Q. J. M., Lally, N., Faulkner, P., Eshel, N., Seifritz, E., Gershman, S. J., Dayan, P., and Roiser, J. P. (2015). Interplay of approximate planning strategies. *Proceedings of the National Academy of Sciences*, 112(10):3098–3103.

Icard, T. (2014a). Toward boundedly rational analysis. Proceedings of the 36th Annual Conference of the Cognitive Science Society.

Icard, T. F. (2014b). *The Algorithmic Mind: A Study of Inference in Action*. PhD thesis.

Jachimowicz, J. M., Duncan, S., Weber, E. U., and Johnson, E. J. (2019). When and why defaults influence decisions: A meta-analysis of default effects. *Behavioural Public Policy*, 3(2):159–186.

Jain, Y. R., Callaway, F., Griffiths, T. L., Dayan, P., He, R., Krueger, P. M., and Lieder, F. (2022). A computational process-tracing method for measuring people's planning strategies and how they change over time. *Behavior Research Methods*.

Jain, Y. R., Gupta, S., Rakesh, V., Dayan, P., Callaway, F., and Lieder, F. (2019). How do people learn how to plan? Conference on Cognitive Computational Neuroscience.

Jaynes, E. T. (2003). *Probability Theory: The Logic of Science*. Cambridge University Press.

Jeffrey, R. C. (1965). *The Logic of Decision*. McGraw Hill.

Johnson, E. J. and Goldstein, D. (2003). Do defaults save lives?

Johnson, E. J. and Payne, J. W. (1985). Effort and accuracy in choice. *Management Science*, 31(4):395–414.

Jones, M. and Love, B. C. (2011). Bayesian fundamentalism or Enlightenment? On the explanatory status and theoretical contributions of Bayesian models of cognition. *Behavioral and Brain Sciences*, 34(4):169–188.

Jones, P. E. and Roelofsma, P. H. (2000). The potential for social contextual and group biases in team decision-making: Biases, conditions and psychological mechanisms. *Ergonomics*, 43(8):1129–1152.

Jung, J. Y. and Mellers, B. A. (2016). American attitudes toward nudges. *Judgment and Decision Making*, 11(1):62–74.

Kaelbling, L., Littman, M., and Cassandra, A. (1998). Planning and acting in partially observable stochastic domains. *Artificial Intelligence*, 101(1–2):99–134.

Kahneman, D. (2011). *Thinking, Fast and Slow*. Farrar, Strauss and Giroux.

Kahneman, D. and Frederick, S. (2002). Representativeness revisited: Attribute substitution in intuitive judgment. In T. Gilovich, D. Griffin, and D. Kahneman (Eds.), Heuristics and Biases: The Psychology of Intuitive Judgment. Pages 49–81. Cambridge University Press.

Kahneman, D. and Frederick, S. (2005). A model of heuristic judgment. Pages 267–293. Cambridge University Press.

Kahneman, D. and Klein, G. (2009). Conditions for intuitive expertise: A failure to disagree. *American Psychologist*, 64(6):515.

Kahneman, D., Slovic, P., and Tversky, A. (1982). *Judgment under Uncertainty: Heuristics and Biases*. Cambridge University Press.

Kahneman, D. and Tversky, A. (1972). Subjective probability: A judgment of representativeness. *Cognitive Psychology*, 3:430–454.

Kahneman, D. and Tversky, A. (1979). Prospect theory: An analysis of decision under risk. *Econometrica*, 47(2):263–292.

Kahneman, D. and Tversky, A. (1996). On the reality of cognitive illusions. *Psychological Review*, 103:582–59.

Kaiser, M., Bernauer, M., Sunstein, C. R., and Reisch, L. A. (2020). The power of green defaults: The impact of regional variation of opt-out tariffs on green energy demand in Germany. *Ecological Economics*, 174:106685.

Kasneci, E., Seßler, K., Küchemann, S., Bannert, M., Dementieva, D., Fischer, F., Gasser, U., Groh, G., Günnemann, S., Hüllermeier, E., et al. (2023). ChatGPT for good? On opportunities and challenges of large language models for education. *Learning and Individual Differences*, 103:102274.

Kassirer, S., Levine, E. E., and Kouchaki, M. (2023). A call to maximize the social impact of our research: An effective altruism approach. *Academy of Management Perspectives*, 37(4):371–382.

Keeney, R. L. (1982). Decision analysis: An overview. *Operations Research*, 30(5):803–838.

Keramati, M., Dezfouli, A., and Piray, P. (2011). Speed/accuracy trade-off between the habitual and the goal-directed processes. *PLoS Computational Biology*, 7(5):e1002055.

Keramati, M., Smittenaar, P., Dolan, R. J., and Dayan, P. (2016). Adaptive integration of habits into depth-limited planning defines a habitual-goal–directed spectrum. *Proceedings of the National Academy of Sciences*, 113(45):12868–12873.

Kerr, N. L. and Tindale, R. S. (2004). Group performance and decision making. Annual Review of Psychology, 55:623–655.

Keynes, J. M. (1936). *The General Theory of Employment, Interest and Money*. Palgrave MacMillan.

Khaw, M. W., Li, Z., and Woodford, M. (2021). Cognitive imprecision and small-stakes risk aversion. *The Review of Economic Studies*, 88(4):1979–2013.

Kinnier, R. T. and Metha, A. T. (1989). Regrets and priorities at three stages of life. *Counseling and Values*, 33(3):182–193.

Klingsieck, K. B. (2013). Procrastination: When good things don't come to those who wait. *European Psychologist*, 18(1), 24–34.

Koedinger, K. R., Anderson, J. R., Hadley, W. H., and Mark, M. A. (1997). Intelligent tutoring goes to school in the big city. *International Journal of Artificial Intelligence in Education*, 8:30–43.

Kollock, P. (1998). Social dilemmas: The anatomy of cooperation. *Annual Review of Sociology*, 24(1):183–214.

Kool, W. and Botvinick, M. (2018). Mental labour. *Nature Human Behaviour*, 2(12):899–908.

Kool, W., Gershman, S. J., and Cushman, F. A. (2017). Cost-benefit arbitration between multiple reinforcement-learning systems. *Psychological Science*, 28(9):1321–1333.

Koriat, A. (1993). How do we know that we know? The accessibility model of the feeling of knowing. *Psychological Review*, 100(4):609–639.

Koster, R., Balaguer, J., Tacchetti, A., Weinstein, A., Zhu, T., Hauser, O., Williams, D., Campbell-Gillingham, L., Thacker, P., Botvinick, M., et al. (2022). Human-centred mechanism design with democratic AI. *Nature Human Behaviour*, 6(10):1398–1407.

Krajbich, I., Armel, C., and Rangel, A. (2010). Visual fixations and the computation and comparison of value in simple choice. *Nature Neuroscience*, 13(10):1292–1298.

Krajbich, I. and Rangel, A. (2011). Multialternative drift-diffusion model predicts the relationship between visual fixations and choice in value-based decisions. *Proceedings of the National Academy of Sciences*, 108(33):13852–13857.

Krueger, P. M., Callaway, F., Gul, S., Griffiths, T., and Lieder, F. (2024). Identifying resource-rational heuristics for risky choice. *Psychological Review*, 131(4):905–951.

Krueger, P. M., Lieder, F., and Griffiths, T. L. (2017). Enhancing metacognitive reinforcement learning using reward structures and feedback. Cognitive Science Society. Proceedings of the 39th Annual Conference of the Cognitive Science Society.

Krusche, M. J. F., Schulz, E., Guez, A., and Speekenbrink, M. (2018). Adaptive planning in human search. Proceedings of the 40th Annual Meeting of the Cognitive Science Society.

Kugler, T., Kausel, E. E., and Kocher, M. G. (2012). Are groups more rational than individuals? A review of interactive decision making in groups. *Wiley Interdisciplinary Reviews: Cognitive Science*, 3(4):471–482.

Kulik, J. A. and Fletcher, J. D. (2016). Effectiveness of intelligent tutoring systems: A meta-analytic review. *Review of Educational Research*, 86(1):42–78.

Kumar, S. and Goyal, N. (2015). Behavioural biases in investment decision making—A systematic literature review. *Qualitative Research in Financial Markets*, 7(1):88–108.

Kurzban, R., Duckworth, A., Kable, J. W., and Myers, J. (2013). An opportunity cost model of subjective effort and task performance. *The Behavioral and Brain Sciences*, 36(6):10.1017/S0140525X12003196, pages 661–679.

Lachman, J. L., Lachman, R., and Thronesbery, C. (1979). Metamemory through the adult life span. *Developmental Psychology*, 15(5):543–551.

Laffont, J.-J. and Martimort, D. (2002). *The Theory of Incentives: The Principal-Agent Model*. Princeton University Press.

Laird, J., Rosenbloom, P., and Newell, A. (1987). SOAR: An architecture for general intelligence. *Artificial Intelligence*, 33(1):1–64.

Lakatos, I. (1970). Falsification and the methodology of scientific research programmes. In I. Lakatos and A. Musgrave (Eds.), *Criticism and the Growth of Knowledge*. Cambridge University Press.

Larrick, R. P. (2004). Debiasing. *Blackwell Handbook of Judgment and Decision Making*, pages 316–338.

Lee, D., Seo, H., and Jung, M. W. (2012). Neural basis of reinforcement learning and decision making. *Annual Review of Neuroscience*, 35(1):287–308.

Levine, S., Chater, N., Tenenbaum, J., and Cushman, F. A. (2023). Resource-rational contractualism: A triple theory of moral cognition. Behavioral and Brain Sciences, doi:10.1017/S0140525X24001067.

Levy, R., Reali, F., and Griffiths, T. L. (2009). Modeling the effects of memory on human online sentence processing with particle filters. *Advances in Neural Information Processing Systems* 21.

Lewis, R. L., Howes, A., and Singh, S. (2014). Computational rationality: Linking mechanism and behavior through bounded utility maximization. *Topics in Cognitive Science*, 6(2):279–311.

Lieder, F. (2018a). *Beyond Bounded Rationality: Reverse-Engineering and Enhancing Human Intelligence*. PhD thesis.

Lieder, F. (2018b). Chapter 4: A rational solution to the strategy selection problem. In *Beyond Bounded Rationality: Reverse-Engineering and Enhancing Human Intelligence*, pages 160–211. University of California, Berkeley.

Lieder, F., Callaway, F., and Griffiths, T. (2024a). *Resource-Rational Analysis*. In T. L. Griffiths, N. Chater, and J. Tenenbaum (Eds.), *Bayesian Models of Cognition: Reverse Engineering the Mind* (pp. 315–340). MIT Press.

Lieder, F., Chen, O. X., Krueger, P. M., and Griffiths, T. L. (2019). Cognitive prostheses for goal achievement. *Nature Human Behaviour*, 3(10):1096–1106.

Lieder, F., Chen, P.-Z., Prentice, M., Amo, V., Tošić, M., et al. (2024b). Gamification of behavior change: Mathematical principle and proof-of-concept study. *JMIR Serious Games*, 12(1):e43078.

Lieder, F., Chen, P.-Z., Stojcheski, J., Consul, S., and Pammer-Schindler, V. (2022a). A cautionary tale about AI-generated goal suggestions. *Mensch und Computer 2022*.

Lieder, F. and Griffiths, T. L. (2015). When to use which heuristic: A rational solution to the strategy selection problem. Volume 1, pages 1–6. Cognitive Science Society.

Lieder, F. and Griffiths, T. L. (2017). Strategy selection as rational metareasoning. *Psychological Review*, 124(6):762–794.

Lieder, F. and Griffiths, T. L. (2020a). Advancing rational analysis to the algorithmic level. *Behavioral and Brain Sciences*, 43.

Lieder, F. and Griffiths, T. L. (2020b). Resource-rational analysis: Understanding human cognition as the optimal use of limited computational resources. *Behavioral and Brain Sciences*, 43:e1.

Lieder, F., Griffiths, T. L., and Goodman, N. D. (2012). Burn-in, bias, and the rationality of anchoring. *Advances in Neural Information Processing Systems* 25.

Lieder, F., Griffiths, T. L., and Hsu, M. (2018a). Overrepresentation of extreme events in decision making reflects rational use of cognitive resources. *Psychological Review*, 125(1):1–32.

Lieder, F., Griffiths, T. L., M. Huys, Q. J., and Goodman, N. D. (2017a). Testing models of anchoring and adjustment. PsyArXiv preprint.

Lieder, F., Griffiths, T. L., Huys, Q. J. M., and Goodman, N. D. (2018b). The anchoring bias reflects rational use of cognitive resources. *Psychonomic Bulletin & Review*, 25(1):322–349.

Lieder, F., Griffiths, T. L., Huys, Q. J. M., and Goodman, N. D. (2018c). Empirical evidence for resource-rational anchoring and adjustment. *Psychonomic Bulletin & Review*.

Lieder, F., Hsu, M., and Griffiths, T. L. (2014a). The high availability of extreme events serves resource-rational decision-making. *Proceedings of the 36th Annual Conference of the Cognitive Science Society*.

Lieder, F., Krueger, P. M., and Griffiths, T. L. (2017b). An automatic method for discovering rational heuristics for risky choice. *Proceedings of the 39th Annual Conference of the Cognitive Science Society.*

Lieder, F., Plunkett, D., Hamrick, J. B., Russell, S. J., Hay, N., and Griffiths, T. (2014b). Algorithm selection by rational metareasoning as a model of human strategy selection. *Advances in Neural Information Processing Systems* 27.

Lieder, F., Prentice, M., and Corwin-Renner, E. (2022b). An interdisciplinary synthesis of research on understanding and promoting well-doing. *Social and Personality Psychology Compass.*

Lieder, F., Shenhav, A., Musslick, S., and Griffiths, T. L. (2018d). Rational Metareasoning and the Plasticity of Cognitive Control. *PLoS Computational Biology*, 14(4):e1006043.

Liu, S., Wright, A. P., Patterson, B. L., Wanderer, J. P., Turer, R. W., Nelson, S. D., McCoy, A. B., Sittig, D. F., and Wright, A. (2023). Using AI-generated suggestions from ChatGPT to optimize clinical decision support. *Journal of the American Medical Informatics Association*, 30(7):1237–1245.

Lu, Q., Hasson, U., and Norman, K. A. (2022). A neural network model of when to retrieve and encode episodic memories. *eLife*, 11:e74445.

Ludvig, E. A., Madan, C. R., and Spetch, M. L. (2014). Extreme outcomes sway risky decisions from experience. *Journal of Behavioral Decision Making*, 27(2):146–156.

Maass, W. (2000). On the computational power of winner-take-all. *Neural Computation*, 12(11):2519–2535.

MacGregor, J. N., Ormerod, T. C., and Chronicle, E. P. (2001). Information processing and insight: A process model of performance on the nine-dot and related problems. *Journal of Experimental Psychology: Learning, Memory, and Cognition*, 27(1):176–201.

Madan, C. R., Ludvig, E. A., and Spetch, M. L. (2014). Remembering the best and worst of times: Memories for extreme outcomes bias risky decisions. *Psychonomic Bulletin & Review*, 21(3):629–636.

Maier, M., Bartoš, F., Stanley, T., Shanks, D. R., Harris, A. J., and Wagenmakers, E.-J. (2022). No evidence for nudging after adjusting for publication bias. *Proceedings of the National Academy of Sciences*, 119(31):e2200300119.

Maisto, D., Donnarumma, F., and Pezzulo, G. (2015). Divide et impera: Subgoaling reduces the complexity of probabilistic inference and problem solving. *Journal of the Royal Society Interface*, 12(104):20141335.

Marakas, G. M. (2003). *Decision Support Systems in the 21st Century*, volume 134. Prentice Hall Upper Saddle River, NJ.

Marcus, G. F. (2008). *Kluge: The Haphazard Evolution of the Human Mind.* Houghton Mifflin Harcourt.

Marr, D. (1982). *Vision.* W. H. Freeman.

Martignon, L. and Hoffrage, U. (1999). Why does one-reason decision making work? A case study in ecological rationality. In Gigerenzer, G. and Todd, P. (Eds.) *Simple Heuristics That Make US Smart*, pages 75–95. Oxford University Press.

Matějka, F. and McKay, A. (2015). Rational inattention to discrete choices: A new foundation for the multinomial logit model. *American Economic Review*, 105(1):272–298.

Matheson, J. E. (1968). The economic value of analysis and computation. *IEEE Transactions on Systems Science and Cybernetics*, 4(3):325–332.

Mattar, M. G. and Lengyel, M. (2022). Planning in the brain. *Neuron*, 110(6):914–934.

May, B. C., Korda, N., Lee, A., and Leslie, D. S. (2012). Optimistic Bayesian sampling in contextual-bandit problems. *Journal of Machine Learning Research*, 13:2069–2106.

McFadden, D. (2001). Economic choices. *American Economic Review*, 91(3):351–378.

Mehta, A., Jain, Y. R., Kemtur, A., Stojcheski, J., Consul, S., Tošić, M., and Lieder, F. (2022). Leveraging machine learning to automatically derive robust decision strategies from imperfect knowledge of the real world. *Computational Brain & Behavior*, 5(3): 343–377.

Mengersen, K. L. and Tweedie, R. L. (1996). Rates of convergence of the Hastings and Metropolis algorithms. *The Annals of Statistics*, 24(1):101–121.

Milkman, K. L., Ellis, S. F., Gromet, D. M., Jung, Y., Luscher, A. S., Mobarak, R. S., Paxson, M. K., Silvera Zumaran, R. A., Kuan, R., Berman, R., et al. (2024). Megastudy shows that reminders boost vaccination but adding free rides does not. *Nature*, 631:179–188.

Milkman, K. L., Gandhi, L., Patel, M. S., Graci, H. N., Gromet, D. M., Ho, H., Kay, J. S., Lee, T. W., Rothschild, J., Bogard, J. E., et al. (2022). A 680,000-person megastudy of nudges to encourage vaccination in pharmacies. *Proceedings of the National Academy of Sciences*, 119(6):e2115126119.

Milkman, K. L., Patel, M. S., Gandhi, L., Graci, H. N., Gromet, D. M., Ho, H., Kay, J. S., Lee, T. W., Akinola, M., Beshears, J., et al. (2021). A megastudy of text-based nudges encouraging patients to get vaccinated at an upcoming doctor's appointment. *Proceedings of the National Academy of Sciences*, 118(20):e2101165118.

Mill, J. S. (1998). *Utilitarianism*. Oxford University Press, New York. Originally published in 1861.

Miller, G. A. (2003). The cognitive revolution: A historical perspective. *Trends in Cognitive Sciences*, 7(3):141–144.

Miller, K. J. and Venditto, S. J. C. (2021). Multi-step planning in the brain. *Current Opinion in Behavioral Sciences*, 38:29–39.

Milli, S., Lieder, F., and Griffiths, T. L. (2021). A rational reinterpretation of dual-process theories. *Cognition*, 217:104881.

Milosavljevic, M., Malmaud, J., and Huth, A. (2010). The drift diffusion model can account for the accuracy and reaction time of value-based choices under high and low time pressure. 5(6):437–449.

Minaee, S., Mikolov, T., Nikzad, N., Chenaghlu, M., Socher, R., Amatriain, X., and Gao, J. (2024). Large language models: A survey. arXiv preprint *arXiv:2402.06196*.

Miranda, S. M. (1994). Avoidance of groupthink. *Small Group Research*, 25:105 – 136.

Mnih, V., Kavukcuoglu, K., Silver, D., Rusu, A. A., Veness, J., Bellemare, M. G., Graves, A., Riedmiller, M., Fidjeland, A. K., Ostrovski, G., Petersen, S., Beattie, C., Sadik, A., Antonoglou, I., King, H., Kumaran, D., Wierstra, D., Legg, S., and Hassabis, D. (2015). Human-level control through deep reinforcement learning. *Nature*, 518(7540):529–533.

Molden, D. C., Hui, C. M., Scholer, A. A., Meier, B. P., Noreen, E. E., D'Agostino, P. R., and Martin, V. (2012). Motivational versus metabolic effects of carbohydrates on self-control. *Psychological Science*, 23(10):1137–1144.

Moreno-Bote, R., Knill, D. C., and Pouget, A. (2011). Bayesian sampling in visual perception. *Proceedings of the National Academy of Sciences*, 108(30):12491–12496.

Musslick, S., Shenhav, A., Botvinick, M. M., and Cohen, J. D. (2015). A computational model of control allocation based on the expected value of control. *Reinforcement Learning and Decision Making Conference*.

Neal, R. M. (1993). Probabilistic inference using Markov chain Monte Carlo methods. Technical Report CRG-TR-93-1.

Nelson, T. O. (1984). A comparison of current measures of the accuracy of feeling-of-knowing predictions. *Psychological Bulletin*, 95(1):109–133.

Nelson, T. O. and Narens, L. (1990). Metamemory: A Theoretical Framework and New Findings. In G. H. Bower (Ed.), The Psychology of Learning and Motivation (Vol. 26, pp. 125–173). Academic Press.

Newell, A. (1990). *Unified Theories of Cognition*. Harvard University Press.

Newell, A. and Simon, H. (1956). The logic theory machine: A complex information processing system. *IRE Transactions on Information Theory*, IT-2:61–79.

Newell, A., Simon, H. A., et al. (1972). *Human Problem Solving*, volume 104. Prentice-Hall.

Nhouyvanisvong, A. and Reder, L. M. (1998). Rapid feeling-of-knowing: A strategy selection mechanism. In V. Y. Yzerbyt, G. Lories, & B. Dardenne (Eds.), Metacognition: Cognitive and social dimensions. Pages 35–52. Sage Publications, Inc.

Niv, Y., Daw, N. D., Joel, D., and Dayan, P. (2007). Tonic dopamine: Opportunity costs and the control of response vigor. *Psychopharmacology*, 191(3):507–520.

Nkambou, R., Mizoguchi, R., and Bourdeau, J. (2010). *Advances in Intelligent Tutoring Systems*, volume 308. Springer Science & Business Media.

Nutt, P. C. (1993). The formulation processes and tactics used in organizational decision making. *Organization Science*, 4(2):226–251.

Nye, B. D., Mee, D., and Core, M. G. (2023). Generative large language models for dialog-based tutoring: An early consideration of opportunities and concerns. *Proceedings of the Workshop on Empowering Education with LLMs—The Next-Gen Interface and Content Generation 2023* (AIEDLLM 2023).

Oaksford, M. and Chater, N. (1994). A rational analysis of the selection task as optimal data selection. *Psychological Review*, 101(4):608–631.

O'Connor, A. M., Rostom, A., Fiset, V., Tetroe, J., Entwistle, V., Llewellyn-Thomas, H., Holmes-Rovner, M., Barry, M., and Jones, J. (1999). Decision aids for patients facing health treatment or screening decisions: Systematic review. *BMJ*, 319(7212):731–734.

O'Donoghue, T. and Rabin, M. (2015). Present bias: Lessons learned and to be learned. *American Economic Review*, 105(5):273–279.

O'Donoghue, T. and Sprenger, C. (2018). Reference-dependent preferences. Volume 1, pages 1–77. Elsevier.

O'Reilly, R. C. and Frank, M. J. (2006). Making working memory work: A computational model of learning in the prefrontal cortex and basal ganglia. *Neural Computation*, 18(2):283–328.

Ortega, P. A. and Braun, D. A. (2013). Thermodynamics as a theory of decision-making with information-processing costs. *Proceedings of the Royal Society A: Mathematical, Physical and Engineering Sciences*, 469(2153):20120683.

Oster, M., Douglas, R., and Liu, S.-C. (2009). Computation with spikes in a winner-take-all network. *Neural Computation*, 21(9):2437–2465.

Oud, B., Krajbich, I., Miller, K., Cheong, J. H., Botvinick, M., and Fehr, E. (2016). Irrational time allocation in decision-making. *Proceedings of the Royal Society B: Biological Sciences*, 283(1822):20151439.

Oulasvirta, A., Jokinen, J. P. P., and Howes, A. (2022). Computational rationality as a theory of interaction. Proceedings of the 2022 CHI Conference on Human Factors in Computing Systems.

O'Doherty, J. P., Lee, S. W., and McNamee, D. (2015). The structure of reinforcement-learning mechanisms in the human brain. *Current Opinion in Behavioral Sciences*, 1:94–100.

Pascal, B. (1702). *Pensées de M. Pascal sur la religion et sur quelques autres sujets, qui ont été trouvées après sa mort parmi ses papiers*. Desprez.

Patrick, J. and Ahmed, A. (2014). Facilitating representation change in insight problems through training. *Journal of Experimental Psychology: Learning, Memory, and Cognition*, 40(2):532.

Pauly, R., Heindrich, L., Amo, V., and Lieder, F. (2022). What to learn next? Aligning gamification rewards to long-term goals using reinforcement learning. Proceedings of the Multi-disciplinary Conference on Reinforcement Learning and Decision Making.

Pauly, R., Kukamjad, T., Heindrich, L., Amo, V., and Lieder, F. (2025). Optimal gamification of self-directed learning: A computational method and its real-world evaluation in an app for learning English. Preprint. https://osf.io/preprints/psyarxiv/s4kqc_v1/.

Payne, J. W. (1982). Contingent decision behavior. *Psychological Bulletin*, 92(2):382–402.

Payne, J. W., Bettman, J. R., and Johnson, E. J. (1988). Adaptive strategy selection in decision making. *Journal of Experimental Psychology: Learning, Memory, and Cognition*, 14(3):534–552.

Payne, J. W., Bettman, J. R., and Johnson, E. J. (1993). *The Adaptive Decision Maker*. Cambridge University Press.

Phillips, J. and Cushman, F. (2017). Morality constrains the default representation of what is possible. *Proceedings of the National Academy of Sciences*, 114(18):4649–4654.

Phillips, N. D., Neth, H., Woike, J. K., and Gaissmaier, W. (2017). FFTrees: A toolbox to create, visualize, and evaluate fast-and-frugal decision trees. *Judgment and Decision Making*, 12(4):344–368.

Piattelli-Palmarini, M. (1994). *Inevitable Illusions. How Mistakes of Reason Rule Our Minds*. Wiley.

Piriyakulkij, W. T., Langenfeld, C., Le, T. A., and Ellis, K. (2024). Doing experiments and revising rules with natural language and probabilistic reasoning. In *Advances in Neural Information Processing Systems 38*.

Poglitsch, C., Safikhani, S., List, E., and Pirker, J. (2024). XR technologies to enhance the emotional skills of people with autism spectrum disorder: A systematic review. *Computers & Graphics*, page 103942.

Polanía, R., Woodford, M., and Ruff, C. C. (2019). Efficient coding of subjective value. *Nature Neuroscience*, 22(1):134–142.

Posner, M. I. and McLeod, P. (1982). Information processing models-in search of elementary operations. *Annual Review of Psychology*, 33(1):477–514.

Prystawski, B., Mohnert, F., Tošić, M., and Lieder, F. (2022). Resource-rational models of human goal pursuit. *Topics in Cognitive Science*, 14(3):528–549.

Puterman, M. L. (2014). *Markov Decision Processes: Discrete Stochastic Dynamic Programming*. Wiley.

Radulescu, A., van Opheusden, B., Callaway, F., Griffiths, T., and Hillis, J. (2020). Modeling visual search in naturalistic virtual reality environments. *Journal of Vision*, 20(11): 1401.

Ramsey, F. P. (1926/1990). Truth and probability. In Mellor, D. H. (ed.) *F. P. Ramsey: Philosophical Papers*, pages 21–45. Cambridge University Press.

Rashidan, M. A., Na'im Sidek, S., Yusof, H. M., Khalid, M., Dzulkarnain, A. A. A., Ghazali, A. S., Zabidi, S. A. M., and Sidique, F. A. A. (2021). Technology-assisted emotion recognition for autism spectrum disorder (ASD) children: A systematic literature review. *IEEE Access*, 9:33638–33653.

Ratcliff, R. (1978). A theory of memory retrieval. *Psychological Review*, 85(2):59–108.

Ratcliff, R. and McKoon, G. (2008). The diffusion decision model: Theory and data for two-choice decision tasks. *Neural Computation*, 20(4):873–922.

Reder, L. M. (1987). Strategy selection in question answering. *Cognitive Psychology*, 19(1):90–138.

Reder, L. M. and Ritter, F. E. (1992). What determines initial feeling of knowing? Familiarity with question terms, not with the answer. *Journal of Experimental Psychology: Learning, Memory, and Cognition*, 18(3):435.

Regenwetter, M., Grofman, B., Popova, A., Messner, W., Davis-Stober, C. P., and Cavagnaro, D. R. (2009). Behavioural social choice: A status report. *Philosophical Transactions of the Royal Society B: Biological Sciences*, 364(1518):833–843.

Reichman, D., Lieder, F., Bourgin, D. D., Talmon, N., and Griffiths, T. L. (2023). The computational challenges of means selection problems: Network structure of goal systems predicts human performance. *Cognitive Science*, 47(8):e13330.

Rieskamp, J. and Otto, P. E. (2006). SSL: A theory of how people learn to select strategies. *Journal of Experimental Psychology: General*, 135(2):207–236.

Robert, C. P. and Casella, G. (2009). *Introducing Monte Carlo Methods with R*. Springer Science & Business Media.

Rostain, T. (2000). Educating homo economicus: Cautionary notes on the new behavioral law and economics movement. *Law & Society Review*, 34(4):973–1006.

Russell, S. J. (1997). Rationality and intelligence. *Artificial Intelligence*, 94(1–2):57–77.

Russell, S. J. and Subramanian, D. (1995). Provably bounded-optimal agents. *Journal of Artificial Intelligence Research*, 2:575–609.

Russell, S. J. and Wefald, E. (1991a). *Do the Right Thing: Studies in Limited Rationality*. MIT Press.

Russell, S. J. and Wefald, E. (1991b). Principles of metareasoning. *Artificial Intelligence*, 49(1–3):361–395.

Russo, J. E. and Dosher, B. A. (1983). Strategies for multiattribute binary choice. *Journal of Experimental Psychology: Learning, Memory, and Cognition*, 9(4):676.

Saeri, A. K., Slattery, P., Lee, J., Houlden, T., Farr, N., Gelber, R. L., Stone, J., Huuskes, L., Timmons, S., Windle, K., et al. (2023). What works to increase charitable donations? A meta-review with meta-meta-analysis. *VOLUNTAS: International Journal of Voluntary and Nonprofit Organizations*, 34(3):626–642.

Sanborn, A. N., Griffiths, T. L., and Navarro, D. J. (2010). Rational approximations to rational models: Alternative algorithms for category learning. *Psychological Review*, 117(4):1144–1167.

Savage, L. J. (1954). *The Foundations of Statistics*. Wiley.

Schmidt, A. T. and Engelen, B. (2020). The ethics of nudging: An overview. *Philosophy Compass*, 15(4):e12658.

Schultz, W., Dayan, P., and Montague, P. R. (1997). A neural substrate of prediction and reward. *Science*, 275(5306):1593–1599.

Schwartz, B. L. and Metcalfe, J. (2017). Metamemory: An update of critical findings. Pages 423–432. Academic Press.

Sederberg, P. B., Howard, M. W., and Kahana, M. J. (2008). A context-based theory of recency and contiguity in free recall. *Psychological Review*, 115(4):893.

Sen, A. (1986). Social choice theory. *Handbook of Mathematical Economics*, 3:1073–1181.

Sezener, C. E. and Dayan, P. (2020). Static and dynamic values of computation in MCTS. Pages 31–40.

Sezener, C. E., Dezfouli, A., and Keramati, M. (2019). Optimizing the depth and the direction of prospective planning using information values. *PLoS Computational Biology*, 15(3):e1006827.

Shenhav, A., Botvinick, M. M., and Cohen, J. D. (2013). The expected value of control: An integrative theory of anterior cingulate cortex function. *Neuron*, 79(2):217–240.

Shenhav, A., Musslick, S., Lieder, F., Kool, W., Griffiths, T. L., Cohen, J. D., and Botvinick, M. M. (2017). Toward a rational and mechanistic account of mental effort. *Annual Review of Neuroscience*, 40(March):99–124.

Shepard, R. N. (1987). Toward a universal law of generalization for psychological science. *Science*, 237(4820):1317–1323.

Shi, L. and Griffiths, T. (2009). Neural implementation of hierarchical Bayesian inference by importance sampling. *Advances in Neural Information Processing Systems* 22.

Shi, L., Griffiths, T. L., Feldman, N. H., and Sanborn, A. N. (2010). Exemplar models as a mechanism for performing Bayesian inference. *Psychological Bulletin and Review*, 17(4):443–464.

Shrager, J. and Siegler, R. S. (1998). SCADS: A model of children's strategy choices and strategy discoveries. *Psychological Science*, 9(5):405–410.

Shugan, S. M. (1980). The cost of thinking. *Journal of Consumer Research*, 7(2):99–111.

Siegler, R. S. (1988). Strategy choice procedures and the development of multiplication skill. *Journal of Experimental Psychology: General*, 117(3):258–275.

Siegler, R. S. (1999). Strategic development. *Trends in Cognitive Sciences*, 3(11):430–435.

Siegler, R. S. and Jeff, S. (1984). Strategy choices in addition and subtraction: How do children know what to do? In C. Sophian (Ed.) Origins of Cognitive Skills: The 18th Annual Carnegie Mellon Symposium on Cognition. Pages 229–293. Erlbaum.

Siegler, R. S. and Shipley, C. (1995). Variation, selection, and cognitive change. In T. J. Simon & G. S. Halford (Eds.), Developing cognitive competence: New approaches to process modeling. Pages 31–76. Erlbaum.

Simon, H. A. (1955). A behavioral model of rational choice. *The Quarterly Journal of Economics*, 69(1):99–118.

Simon, H. A. (1956). Rational choice and the structure of the environment. *Psychological Review*, 63(2):129–138.

Simon, H. A. (1978). Rationality as process and as product of thought. *The American Economic Review*, 68(2):1–16.

Simon, H. A. (1979). Information Processing Models of Cognition. *Annual Review of Psychology*, 30(1):363–396.

Simon, H. A. (1990). Invariants of Human Behavior. *Annual Review of Psychology*, 41(1):1–20.

Simon, H. A. (2013). *Administrative Behavior*. Simon and Schuster.

Sims, C. A. (1998). Stickiness. *Carnegie-Rochester Conference Series on Public Policy*, 49:317–356.

Sims, C. A. (2003). Implications of rational inattention. *Journal of Monetary Economics*, 50(3):665–690.

Sims, C. R. (2016). Rate–distortion theory and human perception. *Cognition*, 152, 181–198. https://doi.org/10.1016/j.cognition.2016.03.020.

Şimşek, Ö. and Barto, A. G. (2009). Skill characterization based on betweenness. Advances in Neural Information Processing Systems 22.

Şimşek, Ö., Wolfe, A. P., and Barto, A. G. (2005). Identifying useful subgoals in reinforcement learning by local graph partitioning. International Conference on Machine Learning.

Singhi, N., Mohnert, F., Prystawski, B., and Lieder, F. (2023). Toward a normative theory of (self-)management by goal-setting. Proceedings of the 45th Annual Meeting of the Cognitive Science Society.

Sinnott-Armstrong, W. (2023). Consequentialism. *The Stanford Encyclopedia of Philosophy*. Accessed: 2024-09-22.

Skinner, B. F. (1953). *Science and Human Behavior*. Simon and Schuster.

Skinner, J. (2007). Are you sure you're saving enough for retirement? *Journal of Economic Perspectives*, 21(3):59–80.

Skirzyński, J., Becker, F., and Lieder, F. (2021). Automatic discovery of interpretable planning strategies. *Machine Learning*, 110:2641–2683.

Snider, J., Lee, D., Poizner, H., and Gepshtein, S. (2015). Prospective optimization with limited resources. *PLOS Computational Biology*, 11(9):e1004501.

Solway, A., Diuk, C., Córdova, N., Yee, D., Barto, A. G., Niv, Y., and Botvinick, M. M. (2014). Optimal behavioral hierarchy. *PLoS Computational Biology*, 10(8):e1003779.

Srinivas, S. C., He, R., and Lieder, F. (2023). Learning planning strategies without feedback. volume 45.

Stachenfeld, K. L., Botvinick, M. M., and Gershman, S. J. (2017). The hippocampus as a predictive map. *Nature Neuroscience*, 20(11):1643–1653.

Stewart, N., Chater, N., and Brown, G. D. (2006). Decision by sampling. *Cognitive Psychology*, 53(1):1–26.

Suchow, J. W. and Griffiths, T. L. (2016). Deciding to remember: memory maintenance as a Markov decision process. Proceedings of the 38th Annual Meeting of the Cognitive Science Society.

Sumers, T. R., Yao, S., Narasimhan, K., and Griffiths, T. L. (2023). Cognitive architectures for language agents. arXiv preprint *arXiv:2309.02427*.

Summerfield, C. and Tsetsos, K. (2015). Do humans make good decisions? *Trends in Cognitive Sciences*, 19(1):27–34.

Sunstein, C. R. (2000). *Behavioral Law and Economics*. Cambridge University Press.

Sunstein, C. R. (2003). Hazardous heuristics. *The University of Chicago Law Review*, 70(2):751–782.

Sunstein, C. R. (2005). Moral heuristics. *Behavioral and Brain Sciences*, 28(4):531–541.

Sunstein, C. R. (2015). The ethics of nudging. *Yale Journal on Regulation*, 32:413–450.

Sunstein, C. R. (2017). Nudges that fail. *Behavioural Public Policy*, 1(1):4–25.

Sunstein, C. R. and Hastie, R. (2015). *Wiser: Getting beyond Groupthink to Make Groups Smarter*. Harvard Business Press.

Sutherland, S. (1992). *Irrationality: The Enemy Within*. Constable and Company.

Sutton, R. S. and Barto, A. G. (2018). *Reinforcement Learning: An Introduction*. MIT Press.

Sutton, R. S., Precup, D., and Singh, S. (1999). Between MDPs and semi-MDPs: A framework for temporal abstraction in reinforcement learning. *Artificial Intelligence*, 112(1–2):181–211.

Tajima, S., Drugowitsch, J., Patel, N., and Pouget, A. (2019). Optimal policy for multi-alternative decisions. *Nature Neuroscience*, 22(9):1503–1511.

Teerapittayanon, S., McDanel, B., and Kung, H.-T. (2016). Branchynet: Fast inference via early exiting from deep neural networks. In *2016 23rd International Conference on Pattern Recognition (ICPR)*, pages 2464–2469. IEEE.

Tenenbaum, J. B., Kemp, C., Griffiths, T. L., and Goodman, N. (2011). How to grow a mind: Statistics, structure, and abstraction. *Science*, 331(6022):1279–1285.

Thaler, R. H. (2000). From homo economicus to homo sapiens. *Journal of Economic Perspectives*, 14(1):133–141.

Thaler, R. H. (2015). *Misbehaving: The Making of Behavioral Economics*. WW Norton & Company.

Thaler, R. H. and Sunstein, C. R. (2021). *Nudge: The Final Edition*. Yale University Press.

Thirunavukarasu, A. J., Ting, D. S. J., Elangovan, K., Gutierrez, L., Tan, T. F., and Ting, D. S. W. (2023). Large language models in medicine. *Nature Medicine*, 29(8):1930–1940.

Thompson, B., van Opheusden, B., Sumers, T., and Griffiths, T. (2022). Complex cognitive algorithms preserved by selective social learning in experimental populations. *Science*, 376(6588):95–98.

Todd, M., Niv, Y., and Cohen, J. D. (2008). Learning to use working memory in partially observable environments through dopaminergic reinforcement. *Advances in Neural Information Processing Systems* 21.

Tomlin, D., Rand, D. G., Ludvig, E. A., and Cohen, J. D. (2015). The evolution and devolution of cognitive control: The costs of deliberation in a competitive world. *Scientific Reports*, 5(1):11002.

Tomov, M. S., Yagati, S., Kumar, A., Yang, W., and Gershman, S. J. (2020). Discovery of hierarchical representations for efficient planning. *PLoS Computational Biology*, 16(4):e1007594.

Tsetsos, K., Moran, R., Moreland, J., Chater, N., Usher, M., and Summerfield, C. (2016). Economic irrationality is optimal during noisy decision making. *Proceedings of the National Academy of Sciences*, 113(11):3102–3107.

Tversky, A. (1969). Intransitivity of preferences. *Psychological Review*, 76(1):31–48.

Tversky, A. (1972). Elimination by aspects: A theory of choice. *Psychological Review*, 79(4):281–299.

Tversky, A. (1982). Evidential impact of base rates. In Kahneman, D., Slovic, P., and Tversky, A. (eds.) *Judgement under Uncertainty: Heuristics and Biases*. Cambridge University Press.

Tversky, A. and Kahneman, D. (1973). Availability: A heuristic for judging frequency and probability. *Cognitive Psychology*, 5(2):207–232.

Tversky, A. and Kahneman, D. (1974). Judgment under uncertainty: Heuristics and biases. *Science*, 185(4157):1124–1131.

Tversky, A. and Kahneman, D. (1983). Extensional vs. intuitive reasoning: The conjunction fallacy in probability judgment. *Psychological Review*, 90:293–315.

Ullman, T. D., Stuhlmüller, A., Goodman, N. D., and Tenenbaum, J. B. (2018). Learning physical parameters from dynamic scenes. *Cognitive Psychology*, 104:57–82.

Usher, M. and McClelland, J. L. (2001). The time course of perceptual choice: The leaky, competing accumulator model. *Psychological Review*, 108(3):550–592.

Vadillo, M. A., Gold, N., and Osman, M. (2016). The bitter truth about sugar and willpower: The limited evidential value of the glucose model of ego depletion. *Psychological Science*, 27(9):1207–1214.

van den Berg, R. and Ma, W. J. (2018). A resource-rational theory of set size effects in human visual working memory. *eLife*, 7:e34963.

van Lange, P. A., Joireman, J., Parks, C. D., and Van Dijk, E. (2013). The psychology of social dilemmas: A review. *Organizational Behavior and Human Decision Processes*, 120(2):125–141.

van Opheusden, B., Galbiati, G., Bnaya, Z., Li, Y., and Ma, W. J. (2017). A computational model for decision tree search. Proceedings of the 39th Annual Meeting of the Cognitive Science Society.

van Rooij, I. (2008). The tractable cognition thesis. *Cognitive Science*, 32(6):939–984.

Vandemeent, J.-W., Paige, B., Tolpin, D., and Wood, F. (2016). Black-box policy search with probabilistic programs. In *Artificial Intelligence and Statistics*, pages 1195–1204. PMLR.

VanLehn, K. (2011). The relative effectiveness of human tutoring, intelligent tutoring systems, and other tutoring systems. *Educational Psychologist*, 46(4):197–221.

Vavekanand, R., Karttunen, P., Xu, Y., Milani, S., and Li, H. (2024). Large language models in healthcare decision support: A review. npj Digital Medicine, 8:263.

Vélez, N., Christian, B., Hardy, M., Thompson, B. D., and Griffiths, T. L. (2023). How do humans overcome individual computational limitations by working together? *Cognitive Science*, 47(1):e13232.

Vesonder, G. T. and Voss, J. F. (1985). On the ability to predict one's own responses while learning. *Journal of Memory and Language*, 24(3):363–376.

von Neumann, J. and Morgenstern, O. (1944). *Theory of Games and Economic Behavior*. Princeton University Press.

Vul, E., Goodman, N. D., Griffiths, T. L., and Tenenbaum, J. B. (2009). One and done? Optimal decisions from very few samples. *Proceedings of the 31st Annual Meeting of the Cognitive Science Society*. Pages 66–72.

Vul, E., Goodman, N., Griffiths, T. L., and Tenenbaum, J. B. (2014). One and done? Optimal decisions from very few samples. *Cognitive Science*, 38(4):599–637.

Vulkan, N. (2000). An economist's perspective on probability matching. *Journal of Economic Surveys*, 14(1):101–118.

Walliser, B. (2007). *Cognitive Economics*. Springer Science & Business Media.

Wang, Y. and Sloan, F. A. (2018). Present bias and health. *Journal of Risk and Uncertainty*, 57(2):177–198.

Wei, X.-X. and Stocker, A. A. (2017). Lawful relation between perceptual bias and discriminability. *Proceedings of the National Academy of Sciences*, 114(38):10244–10249.

Wilkinson, T. M. (2013). Nudging and manipulation. *Political Studies*, 61(2):341–355.

Wirzberger, M., Lado, A., Prentice, M., Oreshnikov, I., Passy, J.-C., Stock, A., and Lieder, F. (2024). Optimal feedback improves behavioral focus during self-regulated computer-based work. *Scientific Reports*, 14(1):3124.

Woodford, M. (2012). Inattentive valuation and reference-dependent choice. Technical report.

Woodford, M. (2014). Stochastic choice: An optimizing neuroeconomic model. *American Economic Review*, 104(5):495–500.

Wu, J., Luan, S., and Raihani, N. (2022). Reward, punishment, and prosocial behavior: Recent developments and implications. *Current Opinion in Psychology*, 44:117–123.

Xu, L., Wirzberger, M., and Lieder, F. (2019). How should we incentivize learning? An optimal feedback mechanism for educational games and online courses. Proceedings of the 41st Annual Meeting of the Cognitive Science Society.

Yao, S., Yu, D., Zhao, J., Shafran, I., Griffiths, T. L., Cao, Y., and Narasimhan, K. (2023). Tree of thoughts: Deliberate problem solving with large language models. *Advances in Neural Information Processing Systems* 37.

Ying, Z., Callaway, F., Kiyonaga, A., and Mattar, M. G. (2024). Resource-rational encoding of reward information in planning. *Proceedings of the 46th Annual Meeting of the Cognitive Science Society*.

Yonelinas, A. P. (2002). The nature of recollection and familiarity: A review of 30 years of research. *Journal of Memory and Language*, 46:441–517.

Zabaras, N. (2010). Importance sampling. Technical report.

Zacks, J. M. and Tversky, B. (2001). Event structure in perception and conception. *Psychological Bulletin*, 127:3–21.

Zhang, Q., Griffiths, T. L., and Norman, K. A. (2023). Optimal policies for free recall. *Psychological Review* 130(4):1104–1124.

Zhao, B., Lucas, C. G., and Bramley, N. R. (2024). A model of conceptual bootstrapping in human cognition. *Nature Human Behaviour*, 8(1):125–136.

Zhao, W. X., Zhou, K., Li, J., Tang, T., Wang, X., Hou, Y., Min, Y., Zhang, B., Zhang, J., Dong, Z., et al. (2023). A survey of large language models. arXiv preprint *arXiv:2303.18223*.

Zhu, J. Q., Sanborn, A. N., and Chater, N. (2020). The Bayesian sampler: Generic Bayesian inference causes incoherence in human probability judgments. *Psychological Review*, 127(5):719–748.

INDEX

Page numbers in italics indicate figures and tables.

ACT-R, cognitive architecture, 111, 137, 168

Adaptive Character of Thought, The (Anderson), 27, 29

AI. *See* artificial intelligence (AI)

algorithm(s), 2, 8, 12, 48, 68, 140; AI-Interpret, *162*, 162–63, *163*; importance sampling, 75–76; learning, 26, 206n1; Metropolis-Hastings algorithm, 67; planning, 142–44; sampling, 31, 65–66; SARSA, 95

algorithmic level, rational process models, 30

Allais paradox, 22, 24, 25, 64, 83

anchoring-and-adjustment, 71, 72; heuristic, 14, 64; process, 24; resource-rational, 68, 69, 70

anchoring bias, 71, 72, 73; experiments, 73, 74; resource-rational perspective on, 64–74

Anderson, John: *The Adaptive Character of Thought*, 27, 29; rational analysis, 60

architecture(s), 140; definition, 61; resource-rational cognitive architectures, 146–48

artificial intelligence (AI), 2, 19, 67, 69; implications of resource-rational analysis for AI, 200–201

attention, 31–32, 75–83, 125–28

availability biases, resource-rational perspective, 74–83

availability heuristic, 14, 24

axiom(s) of rational choice, 17, 18, 84

backward induction, 224–26; recursive implementation, 225

Bayesian approach to cognitive science, 215

Bayesian belief updating, 127, 216–18, 221

Bayesian decision theory, 65, 66

Bayesian inference, 28, 30–31, 122, 126, 135, 214

Bayesian learning, 94

Bayesian metalevel MDP(s), 215–19; belief states, 215–16; belief updating, 216; partially observable MDPs (POMDP), 218–19

Bayesian metalevel policy search (BMPS), 228–31; learning to select computations, 230–31; value of information (VOI), 229, 230

Bayesian model(s) of cognition, 13, 14, 30, 94, 215

Bayes' rule, 28, 29

behavioral economics, 172, 197; rise of, 203–4

behavioral law, 198

behaviorism, 3, 4, 42

Bellman equations, 224

Bernoulli, Daniel: expected value, 16; subjective value, 16

Bernoulli, Nicolas: expected value, 16; subjective probability, 17

Bernoulli distribution, 217; metalevel probability model, 217–18

bias(es), 23, 151; rationality, 37. *See also* heuristics and biases

bias-variance trade-off, 10, 26

blinkered approximation, 227–28
BMPS. *See* Bayesian metalevel policy search (BMPS)
boost(s), 157, 159, 167, 169, 185, 186, 189, 194
boosting: decision-making, 156, 157; decision-making with resource-rational heuristics, 159–69; future directions, 167–69; optimal, 160–61; theory of resource-rational, 167
bounded optimality, 2, 11, 14; accounting for cognitive constraints, 43–46; cognitive system, *39*; interpolating between value of cognition (VOC) and, 53–57; resource-rational analysis and, 32–33
bounded rationality, 19–21, 33–34; Simon on, 8–9
brain points, 181

ChatGPT, 170
choice architecture, 152–59, 172–85, optimal nudging as framework for choice architecture, 178–80
classification, 168, 200
cognition: Bayesian models of, 30; benefit of, 47; cost of, 49–53; human, 2, 3, 19, 22, 27, 29, 30, 33, 36, 59, 63, 88, 156, 185, 233; as resource, 59; social, 170; value of, *39*
cognitive architecture(s), 43, 54, 61; ACT-R, 61, 111, 137, 168; metalevel MDP, *115*; SOAR, 61, 111, 137
cognitive augmentation, *188*; approach, 169; cognitive prostheses, 170; decision-making, 155, 157, 169–72
cognitive control, plasticity of, 108–9
cognitive costs, metalevel MDP, *129, 209*
cognitive mechanism, 5; rationality, 13–14
cognitive operation(s), 48; metalevel MDP, *115*, 116, 122, 127, 130, 135
cognitive policies, 44
cognitive process(es), 5, 43, 55; decision-making strategies, 111–12; general

framework for characterizing optimal, 136–37; optimal sequential, 213
cognitive prostheses, 3–4, 170, 171
cognitive psychology, 6, 141, 149
cognitive resources, 2, 3, 5, 15, 50, 63; bounded, 6
cognitive science, 11, 13
cognitive strategies: discovering, 201–2; learning better, 12
cognitive systems, resource rational cognitive systems, 56
cognitive training, decision-making, 164–67
cognitive tutor(s), decision-making, 164–67
collective decision-making, 203
compensatory, 106
computation(s), 48, 200; metalevel MDP, *209*, 210–11; views of, 214–15
computational architecture C, 43, 69
computational cost. *See* cognitive costs
computational level, 30, 61
computational rationality, term, 11, 33
computer science, 8, 14, 140
conjunction fallacy, representativeness heuristic, 23
constrained optimization, bounded rationality, 9
construal(s): definition, 146; efficient planning, 144–46; navigation problem, 144–46, *145*
continuity, axiom, 18
cost: metareasoning, 148, *149*; opportunity, 50
cost-benefit trade-offs, 46
cost function, metalevel MDP, *115*, 116, 122, 127, 130
cost of cognition, 49–53, 58
Cox's theorem, 28
credit assignment problem, 228

debiasing, decision-making, 155
decision-making, 2, 151–52; approaches to improving, 152–56; beyond cognitive

strategies, 193–94; beyond individual, 190–93; beyond rational self-interest, 186–87, 189–90; boosting, 156; boosting, with resource-rational heuristics, 159–69; cognitive augmentation, 155, 169–72; cognitive tutors and training, 164–67; collective, 203; debiasing, 155; evidence for human irrationality, 3; formalizing rational, 17–19; future work on resource-rational approaches, 188; game theory, 13n1; generating decision aids, 162–63; heuristics, 14–15; human, 5–6; human behavior, 12–13; human rationality and, 109–10; incentives, 154; learning how to decide, 99–102; learning to become more rational, 102–9; metacognitive feedback, 164, 165; neuroscience, 198–200; nudging, 154–55, 183–84; optimal boosting, 160–61; optimal incentives, 175; optimal incentive structures, 172–74; prospect theory, 24; providing decision support, 155; resource rationality, 156–59; resource-rational nudging, 176–83; restricting choice to good options, 153–54; risk aversion, 85–86; teaching people better decision strategies, 155–56

decision mechanisms: model-based, 48; model-free, 48

decision problem(s), 26, 38, 89, 99, 101, 106, 111, 112, 113

decision procedures, 21, 76

decision process(es), 191

decision science, resource-rational analysis and, 196–97

decision support, 153, 155

decision tree: search, 130n4; thinking ahead, 128

de Montmort, Pierre, on expected value, 16

directed cognition model, 227

Donders, Franciscus, work on "mental chronometry," 111

"drift rate," 48

"Dutch book" argument, 28

EBA. See Elimination-By-Aspects (EBA) heuristic

ecological rationality, 25–27, 34, 106; Gigerenzer's theory of, 109; notion of, 10

economics, 1,7, 14, 196; optimal incentive design, 172; rational inattention in, 10–11; resource-rational analysis, 197–98

elementary information processes (EIPs), 111; concept, 210

Elimination-By-Aspects (EBA) heuristic, 99–101

environment, 40, 45

episode, Markov decision process (MDP), 114

error cost, expected value of, 70, 71

evidence accumulation, 125, 137n6

expected utility, 1, 18, 70; action, 74–75, 76; expected opportunity cost and, 77–81; maximizing, 3

expected utility theory (EUT), 38, 39, 196, 197; cognitive system, 39; Markov decision process (MDP), 119, 120; rational system, 38–42

expected value, 16

expected value of control (EVC) theory, 48

expected value theory, 197

fast-and-frugal heuristic(s), 93, 99–101; decision-making, 168; resource-rational analysis, 167

feature vectors (f), 91–92

feedforward neural network, 95, 96

framing, 32, 188, 193, 194

frequency judgments, 25

game theory, 13n1

gamification app, to-do list gamification, 175

Gigerenzer, Gerd: adaptive heuristics, 25; on bounded rationality, 9; ecological rationality, 26; on ecological rationality, 10

global architecture, 54
Google Maps, 170
Google Search, 170
groups: behavior of, 202–3; decision-making, 190–93
groupthink, 190, 191

hash function, 226
heuristic(s), 14, 23; decision-making, 14–15; Elimination-By-Aspects (EBA), 99–101; fast-and-frugal, 99–101; lexicographic heuristic (LEX), 99–101; optimal boosting teaching people, 160–61; Take-The-Best (TTB), 105–8; weighted additive strategy (WADD), 105–8
heuristics and biases: anchoring-and-adjustment, 24; availability heuristic, 24; conjunction fallacy, 23; framework, 34; Kahneman and Tversky's program, 10, 87–88; representativeness heuristic, 23; research program, 23; resource-rational models, 83–87
hierarchical reinforcement learning (HRL), 139, 202
homo economicus model, 173
homunculus, 137–39; term, 138
Horvitz, Eric: building AI systems, 32; term bounded optimality, 43
Howes, Andrew: cognition, 33; resource rationality, 37
human behavior, 2, 7; decision-making, 12–13; model predictions, 62; models of, 13–14; realistic standard, 63. *See also* decision-making
human cognition. *See* cognition
human irrationality, decision-making, 3
human rationality, 3; debate about, 87–88; improving decision-making, 109–10

Icard, Thomas: bounded optimality, 33; resource rationality, 37
implementation level, rational process, 30
importance sampling, 75–76, 78

incentive(s): decision-making, 154, 157; optimal, *188*, 190, 192; optimal structures for decision-makers, 172–74
independence of irrelevant alternatives, axiom, 18
induction, backward, 224–26
inductive bias, rationality, 36
inductive problems, 28
intelligent agent(s). *See* artificial intelligence
intelligent system. *See* artificial intelligence
intelligent tutoring (systems), 166, 168, 169
irrationality, 3, 8, 14, 29, 63–64, 74, 87, 88, 198

Jeffrey, Richard, utility function, 18

Kahneman, Daniel: heuristics and biases research, 10; rationality, 22
Keynes, John Maynard, on making decisions, 17

large language models (LLMs), 168–69, 170
law and policy, human behavior, 198
learning mechanism(s), 91–93, 95–97, 109–10, *165*
learning process, modeling, 138
levels of analysis, Marr's, 30
Lewis, Richard: cognition, 33; resource rationality, 37
LEX. *See* lexicographic heuristic (LEX)
lexicographic heuristic (LEX), 99–101
loss aversion: explaining, 85; phenomenon, 84

machine learning, 67, 69
Markov chain Monte Carlo (MCMC), 31, 67, 69–70, 141
Markov decision process(es) (MDPs): contextualizing, 119–20; illustration, *206*; Markov property, 222–23; mathematical details, 205–8; metalevel, 114–19, 208–12; navigation problem, 145–46;

neuroscience, 199; optimal policies and value functions, 207–8; sequential decision problems, *113*, 113–14

Marr, David, levels of analysis, 30

maximum entropy optimal policy, definition, 208n3

MCMC. *See* Markov chain Monte Carlo (MCMC)

MDP(s). *See* Markov decision process(es) (MDPs)

memoization, 226, 224

memory, 111, feeling of knowing in memory recall, 132–36, *134*

mental chronometry, 111

mental state(s), 57; metalevel MDP, *115*, 116, 122, 126, 130, *129*, *134*, 135, *209*, 210

metacognition, 201

metacognitive feedback, 164, *165*

metacognitive learning, 102, 200

meta-greedy policy, 226–27

metalevel Markov decision processes (MDPs), 15, 113, 114–19, 201; attention in preferential choices, 125–28, *126*; backward induction, 224–26; Bayesian, 215–19; blinkered approximation, 227–28; cognitive costs, *209*; components of, *115*; computations, *209*, 210–11, 214–15; contextualizing, 119–20; external action, *209*; external reward, *209*; feeling of knowing in memory recall, 132–36, *134*; identifying good policies, 223–28; illustration, *209*; mathematical details, 208–12; mental states, *209*, 210; multistep lookahead, 227–28; myopic policy, 226–27; optimal policies for, 213–14, *214*; policies, 213–14; reward function, 211–12; termination operation, *209*; thinking ahead, 128–32, *129*; transition function, 211; value of computation (VOC), 223–24; world states, *209*, 210

metalevel policy, Markov decision policy (MDP), 117

metamemory, 133

metareasoning; cost of, 148, *149*; rational, 140; sequential problem, 118; testing the predictions of rational, model, 97–102

metareasoning model: learning how to decide, 99–102; learning how to sort, 97–99; testing the predictions of rational, 97–102

methodological assumption, resource rationality as, 13

methods: computational, 8, 34, 161, 163, 186; sampling, 66, 75; strategy discovery, *161*, 163, 168

Metropolis-Hastings algorithm, 67

Monte Carlo. *See* Markov Chain Monte Carlo (MCMC)

Morgenstern, Oskar, on making decisions, 17

Mouselab-MDP paradigm, 164

Mouselab paradigm, 124–25, 131; decision-making, *179*; gambling game, 102–3

multiple-cued recall task, *134*, 136

myopic policy, 226–27; definition, 226

navigation problem, construals, 144–46, *145*

neural circuits, 138

neural network(s), feedforward, 95–96, *96*

neuroscience, 7–8, 196; human decision-making, 198–200

Newell, Allen, human cognition, 111

non-compensatory, 105–6; environments, 26

nudge(s). *See* nudging

nudging, 4, 15, 176; decision-making, 154–55, 157; ethics of, 184; future directions, 183–84; optimal, 176–78, *182*; optimal, as framework for choice architectures, 178–80; real-world applications of optimal, 180–83; resource-rational perspective, *179*. *See also* optimal nudging

objective function, 38
opportunity cost, 50, 51, 52
optimal boosting, *188*
optimal cognitive system, 40
optimal incentives, *188*
optimality, 38, 42
optimal nudging, 176–78, *188*; definition, 176; as framework for choice architectures, 178–80; real-world applications of, 180–83. *See also* nudging
optimal policy: Markov decision process (MDP), 114; metalevel MDP, 213–14, *214*
optimal rationality enhancement, 185
optimal stopping, decision-making, 20
optimization, 37, 43; Bayesian, 231; framework, 32–33
optimization problem, 5, 7, 9, 49, 54, 60, 171, 184–85, 201
organizations: behavior of, 202–3; decision-making, 190–93

partially observable Markov decision processes (POMDPs), 218–19
perfect rationality, 41
planning: construals, 144–46; subgoals, 142–44
policy, 41; cognitive system, 41; Markov decision process (MDP), 114; policies as programs, 202
POMDP(s). *See* partially observable Markov decision processes (POMDPs)
posterior distribution, 28
predictably irrational, 25
preference reversals, 84
preferences, 17
present bias, resource-rational analysis, 86
prior distribution, 28
probabilistic inference, 67
probability, 5, 25
probability matching, 5, 66
problem domain, 51; metalevel MDP, *115*, 116

process models, 67, 70, 190, 192; psychological, 76, 88; rational, 30–32, 34, 65, 86, *96*
process-tracing paradigm(s), 62, 121
program π, definition, 43
prosocial, 187, *188*, 189–91, 194
prospect theory, 197; decision-making, 24
psychology, 6–7, 14

Q-learning, 231

Ramsey, Frank, on making decisions, 17
rational, learning to become more, 102–9
rational action, new view of the mind, 203–4
rational altruism, 187–93
rational analysis, 2, 27–30, 34; Anderson's, 27, 30, 60; Bayes' rule, 28; definition, 60; inductive problems, 28; method, 2
rational decision-making, formalizing, 17–19
rational inattention, 7, 10–11, 31–32, 34
rationality, 1; alternative definition of, 2; bounded, 8–9, 19–21, 33–34, 90, 202; bounded optimality, 43–46; classical, 1, 5, 6, 33, 63, 64; cognitive systems, 59–60; computational costs of, 10; cost of cognition, 49–53; expected utility theory (EUT), 38–42; formalizing, 38–60; human, 3, 87–88, 109–10; models of human behavior, 13–14; notions for cognitive systems, 39; rational choice of mental action, 46–49; resource rationality, 13, 36, 37, 39, 53–57, 202, 233; sequential case, 57–59; value of cognition (VOC), 46–49
rationality enhancement, 158, 184–85; decision process, 158, *158*
rational metareasoning, 8, 14, 48, 140; model, 97–102
rational process model(s), 30–31, 34; strategy selection learning, *96*

reasoning, 6, 12, 24, 28, 29, 31, 48, 63, 66, 80; causal, 70; model-based, 228; model-free, 228; resources, 43
reference-dependent preferences, 84
reinforcement learning, 7, 93; hierarchical, 139, 202; model-based, 95; model-free, 78
representation(s), 140; construals, 144–46; improving mental representations, 193–94; navigation problem, 145–46, 147; resource-rational, 141–46; subgoals, 141–44, 143
representativeness heuristic, 23
resource-constrained cognition, 53, 59n6, 136, 139
resource-rational: altruism, 187–93; anchoring-and-adjustment, 68, 69, 70, 73; architectures, 146–48; cognitive policy, 54–55, 118; cognitive tutor teaching planning strategy, 166; illustration of best intervention, 157; learning from experience, 103–8; model of numerical estimation, 73; nudging, 176–78; perspective on anchoring bias, 64–74; representation, 141–46; teaching, heuristics by demonstration, 161–62
resource-rational analysis, 2, 3, 4, 12, 14, 36, 60–62, 140; answering questions, 33–35; artificial intelligence (AI), 200–201; assumptions behind, 63; bounded optimality and, 32–33; cognitive science, 15; collective decision-making, 203; computer scientist, 8; curious person, 8; economics, 197–98; economist, 7; expanding the scope of, 148–50; foundations of decision science, 196–97; interventions for decisions, 15; metalevel Markov decision processes (MDPs), 121–36; neuroscientist, 7–8; optimal incentive structures, 172–74; planning algorithms, 142–44; present bias, 86; psychologist, 6–7; simple example of, 4–6; steps of, 60–61; strategy selection, 14–15; subgoals, 142–44, 143

resource-rational approach, 3
resource-rational cognitive policy, 54
resource-rational framing, 188, 194
resource rationality, 36, 196; cognitive system, 39; definition, 57–58, 58; environment, 56; framework for decision-making enhancements, 156–59; mathematical formalism, 194; methodological assumption, 13; rationality enhancements, 158, 158; theory of, 158–59, 168
resource-rational model(s), 11, 85; understanding heuristics and biases, 83–87
resource-rational perspective: anchoring-and-adjustment, 68, 69, 70, 73; availability biases in judgment and decision-making, 74–83
reward, Markov decision process (MDP), 113
reward function(s): marginal, 220–21; metalevel MDP, 211–12
risk aversion, human decision-making, 85–86
Russell, Stuart: building AI systems, 32; optimization of computational utility, 43

St. Petersburg paradox, 16, 24
sampling, 65–66, 67; importance sampling, 75–76, 78; methods, 4–5; Thompson sampling, 94
SARSA algorithm, 95
satisficing, 20, 124; heuristic, 86; Simon's model of, 9
SAT-TTB, 124
Savage, Leonard, utility function, 18
sequential case, resource rationality, 57–59
sequential decision problem(s), 112; cognition as, 112–20; external environments, 113; Markov decision process (MDP), 113, 113–14; metalevel MDPs, 114–19
Simon, Herbert: bounded rationality, 8–9, 19–21, 202; human cognition, 111; model of satisficing, 9

Sims, Christopher, rationality in economics, 31

SOAR, cognitive architecture model, 111, 137

social choice theory, 192

social dilemmas, 191, 191–92

society, decision-making, 190–93

"softmax" rule, 19, 32

speed-accuracy trade-off, 74

state, Markov decision process (MDP), 113, *113*

state space(s), 201, 224–26

statistics, 1, 67, 69, 71

stimulus-response learning, 95

strategies, 15; cognitive, 12, 98–99; model-based, 228; model-free, 228; optimal, 26, 90–92, 104–5, 162–64, 168; resource-rational, 122–23, 161–62, 164, 167–68, 177; sorting, 97–99; weighted additive strategy (WADD), 99–100

strategy discovery, 108–9, 111, 202; AI-Interpret, *162*; attention in preferential choices, 125–28, *126*; automatic, *162*, *163*, 168, 198; feeling of knowing in memory recall, 132–36, *134*; metalevel MDP, *123*; planning, 128–32, *129*; risky choice, 122–25

strategy selection, 15; computational models, 102; problem, 90–92; rational process model of, *96*, 97

strategy selection learning, 103–8

strategy selection mechanism(s), 96

Stroop task, 108–9

subjective probability, 17

subjective value, 16, 17

Take-The-Best (TTB), 26, 87, 105–8, 111, 124; strategy discovery, 122, *123*

tallying heuristic, 217

teleological explanations, rational model, 37

termination operation, metalevel MDP, *115*, 117

Theory of Games and Economic Behavior (von Neumann and Morgenstern), 17

Thompson sampling, 94

transition function(s): marginal, 219–20; Markov decision process (MDP), 113; metalevel MDP, *115*, 116, 122, 127, 130, 135, 211

transitivity, axiom, 18

TTB. *See* Take-The-Best (TTB)

Tversky, Amos: heuristics and biases research, 10; rationality, 22

"Type II" rationality, 41

unbounded optimality, 60

utility, metalevel MDP, 122

utility function(s), 18, 38; metalevel MDP, *115*, 126, 130, 133

utility-weighted sampling (UWS), 78, 79; model, 80, 81–83

UWS. *See also* utility-weighted sampling (UWS)

value function(s), 24, 207–8, 223–24, 226

value of cognition (VOC), 39, 51; bounded optimality and, 53–57; cognitive system, 39; metalevel MDPs, 119, *120*; rational choice of mental action, 46–49; strategy selection problem, 90–92

value of computation (VOC), 223–24

value of information (VOI), 228, 229; bounding, 229–30; illustration of features, *230*

VOC. *See* value of cognition (VOC); value of computation (VOC)

VOI. *See* value of information (VOI)

von Neumann, John, on making decisions, 17

WADD. *See* weighted additive strategy (WADD)

weighted additive strategy (WADD), 99–100, 105–8, 111, 122, 123, 124

working memory, 138

world state(s), 57; metalevel MDP, *115*, 116, 122, 126, 130, 133, *209*, 210

GPSR Authorized Representative: Easy Access System Europe - Mustamäe tee
50, 10621 Tallinn, Estonia, gpsr.requests@easproject.com

www.ingramcontent.com/pod-product-compliance
Lightning Source LLC
Chambersburg PA
CBHW031236050326
40690CB00007B/829